Experimental and Theoretical Study of Strength and Stability of Soil

Experimental and Theoretical Study of Strength and Stability of Soil

A.Z. Khasanov & Z.A. Khasanov

***Geofundamentproject*, The Samarkand State Institute of Architecture and Civil Engineering, Samarkand, Uzbekistan**

CRC Press is an imprint of the
Taylor & Francis Group, an **informa** business
A BALKEMA BOOK

Translators: Latypova Nargiza and Yodgor Obidov
Designer: Uzakov Mamur

CRC Press/Balkema is an imprint of the Taylor & Francis Group, an informa business

First published: А.З.Хасанов З.А.Хасанов

ЭКСПЕРИМЕНТАЛЬНО-ТЕОРЕТИЧЕСКИЕ ИССЛЕДОВАНИЯ ПРОЧНОСТИ И УСТОЙЧИВОСТИ ГРУНТОВ

ГП Издательство "Zarafshon" Самарканд – 2015

Typeset by Apex CoVantage, LLC

Library of Congress Cataloging-in-Publication Data
Applied for

Published by: CRC Press/Balkema
Schipholweg 107c, 2316 XC Leiden, The Netherlands
e-mail: Pub.NL@taylorandfrancis.com
www.crcpress.com – www.taylorandfrancis.com

ISBN: 978-0-367-36888-3 (Hbk)
ISBN: 978-0-429-35186-0 (eBook)

Contents

Preface

The book presents experimental results on the strength of sand and clay soils under the conditions of the flat shear triaxial stress state and passive and active loading. The authors compare experimental results with existing theories of strength and identify the reasons for discrepancies. They analyze the experimental data to determine the position of the surface shear within the active and passive resistance of the soil. We consider the new concept of the theory of strength of soils, which allows taking into account the basic strength parameters of soil: the angle of internal friction, cohesion, and coefficient of lateral pressure. The analytical equations make it possible to determine the state of stress at the sites rejected with respect to the principal stresses. We present the definition of the physical nature of the concept of lateral pressure for soil coefficient.

This book describes the results of experiments to determine the critical load on the ground in the shape of a truncated cone, where there were experimentally determined surface shear deflection of angles, on the basis of which we determined the analytical equations that predict the critical load. The book contains a variety of methods for solving geotechnical problems using the theory proposed by the authors of soil strength theory.

The book is intended for professionals working in the field of soil mechanics and geotechnical engineering, as well as for master's and doctoral students of engineering and building profile.

Introduction

A brief historical overview of the evolution of the theory of strength of soils

The reliability and durability of all structures, including the ground, depend on the strength and stability of natural and artificially structured soils. In turn, a safe state of stress in a ground massif is a reliable guarantor of the durability of structures. However, to assess properly the state of stress and its impact on the carrying capacity, or the strength, of the soil is a very difficult task. The records of all the factors influencing the complex geological structure of the soil and, in addition, of the lithosphere (subgrade) is not always possible. For this reason, the historical practice in the construction of all types of buildings was first developing practical experience. In the process of historical development, the construction develops gradually increasing complexity of the geometric configuration of structures and the growth of additional loads transmitted to the foundation soil. It became necessary to assess the reliability of accounting for the strength of different types of soils. For example, when building on sand, constructors historically were aware of this as not reliable, especially during earthquakes, and likewise for places where the groundwater level was high. When selecting sites for construction of settlements, in addition to favorable climatic and soil conditions, special attention was paid to safety from natural disasters (strong winds, floods, earthquakes, volcanoes, landslides, etc.). The information on these accumulated through the historical chronology. Analysis of the development of historical cities and the construction of monumental buildings in Asia, Europe and America shows that basically they were built on areas with a relatively elevated terrain. Moreover, as the criterion of reliability and manufacturability of the design, soils were estimated by strength to mechanical stress. There were primitive technologies to improve the properties of soils. Too strong soils (for example, rock) and very soft soils were not recommended as the basis for the construction of monumental buildings. In the presence of soft soils, for example, water-saturated loose sand or clay soils, people began using wooden piles (in the history of European urban planning); the presence of weak porous loess soils or bulk soil compacted layers required treading by horses and camels. There are cases of soil compaction by soaking it for a long period of time (in the history of the building of the Bibi-Khanim Mosque in Samarkand, in the 14th century). Practically no construction took place along the route of the old Sai (mountain river), even during a prolonged absence of water in these areas.

It was considered unreliable to engage in construction on the sides of slopes and ravines and in areas subject to mechanical suffusion. There was a clear understanding of the reliability of structures and the role in it of foundations and the foundation basis.

The foundation made very high demands, and therefore foundations were created of durable natural quarried stone on waterproof mortar. It was believed that the greater the load on the building, the deeper the foundation should be. There was also another criterion: the relation of the building height to the depth of the foundation. Thus, the main criterion for the reliability of structures associated with a soil base is the depth of its inception on a principle: heavy loads – large depth of embedment. Such a fundamental criterion of reliability is preserved in modern construction.

Based on the above, it can be concluded that the reliability of monumental buildings is estimated on the basis of non-deformability of the subgrade load. This principle of construction of buildings is possible only on one occasion when such a distributed load is given to the foundation soil, which provides a large margin of safety.

We should also note the great role of Central Asian scholars who stood at the origins of mechanics and geology. Al-Farabi Abu Nasr Muhammad ibn Tarajal (870–950) invented the first device that could forecast the groundwater level and the water level of the River Nile in Cairo. Abu Raikhan Muhammad ibn Ahmad al-Biruni (973–1048), a Central Asian scholar and lexicographer, authored major works on mechanics, geodesy and geology.

In modern conditions, the principle described above is unacceptable, because of the massive construction of buildings, which increases the levels of stress, increasing the degree of responsibility and risk. Modern engineering science, based on a profound development of the science of mechanics, is able to predict the stress-strain state and the degree of risk of fracture in the body of complex material properties, as well as in complex engineering and the geological environment.

Here we must emphasize the fact that the mechanics of soils is a relatively young science, compared to classical mechanics as a whole. For this reason, at the beginning of the development of soil mechanics to solve engineering problems, to simulate the tensely deformed state and to determine the strength and sustainability, basic laws of classical mechanics and patterns inherent in a solid medium continuity were used.

Development of the theory of elastic-plasticity, ductility and, in general, nonlinear mechanics allowed approximating the predicted calculated results to the actual measurements in nature. Progressive study of the natural properties of soils as a fractured porous medium made it possible to develop a model of soil, allowing more accurate prediction of the tensely deformed state, and the strength and stability of massive geological structures, built of natural sedimentary rock strata.

For example, Mohr's theory was initially intended for graphic interpretation of the definition of the state of stress at any site in a plane problem for continuous media. This handy mathematical apparatus is adapted for use in determining the maximum angle of deflection of pad shear within active and passive loading, and the coefficient of lateral pressure. However, even the naked eye can notice that the angle of deflection of the most dangerous areas, the coefficient of lateral pressure and the angle of internal friction inherent to the soil as the medium with internal friction cannot be combined in a single equation.

For this reason, many authors who conduct experimental studies of soil face some discrepancies in theoretical results with the classical theory of Mohr-Coulomb strength. Many experts still believe that the strength theory of Mohr for the soil is a certain alternative to the theory of Coulomb strength.

It is known that initially the theory of dry friction of Coulomb was used to assess the strength of the soil. This theory simulates an ideal condition for plasticity in shear. However, when modeling the stress state of the array of soil volume, experts faced problems such as the evaluation of its strength under triaxial compression. Research has begun to find the most dangerous planes, on the surface of which the conditions of ultimate equilibrium corresponding to the condition of Coulomb strength would be observed. The scientific solution to this problem is devoted to the works of many scientists in the world. However, until now, when solving engineering problems of geotechnics, for example, when determining soil pressure on fences, specialists are faced with some discrepancies between the calculated and actual results.

In this regard, we would like to emphasize the words of the famous Russian scientist M.V. Malyshev [9] "*. . . But related issues (issues of soil strength), unfortunately, yet still close to their resolution, although our knowledge deepens systematically through accumulating facts and new individual decisions. . . . However, as it usually happens, the deeper our aspirations, the more difficult is getting a significant result. With the deepening in this or that question, the original concept that seemed clear, turned out to be unproven, supplemented, and sometimes changed. This is the natural way of knowing.*"

From Hooke's law of elasticity theory we know that under the plane stress state, the relative lateral deformation is directly proportional to the relative longitudinal deformation. This proportion is estimated by some coefficient, proposed by Poisson. If you solve the equation of state of the theory of elasticity with respect to the voltage, you can get another kind of this coefficient – the coefficient of lateral pressure. Thus, both of these parameters are parameters of the equation of state relating to the theory of the deformation. When evaluating the strength of ground, as shown above, the limit voltage ratio (ratio of lateral pressure), along with the parameters of the angle of internal friction and the cohesion, can be attributed also to the strength parameters. The physical meaning of the coefficient of correlation of limit voltages (coefficient of lateral pressure) is described as the ultimate resistance of the soil to the deviatory stresses.

Traditionally, it is characterized as a critical ratio of principal stresses at which we can observe the destruction of the soil. It is experimentally proved that the magnitude of this factor depends on the state of stress and displacement (shear) of soil at specific sites. The ground is different from other materials in that, when moving, its hardening occurs due to a dense packing of particles in the volume. Unfortunately, these models do not take into account such important properties of the soil.

In this context, some of the basic tenets of the traditional theory of strength of soils that have become familiar in their practical use in solving engineering problems require some rethinking. For instance, this category of issues includes the angle of deflection of the most dangerous area in the plane strain condition; values of the state of stress at the site, including site shear; the physical essence of strength parameters coefficient of lateral pressure at rest and limit stress. This book attempts to give a slightly different interpretation of the physical nature of these fundamental strength parameters of soils.

Material presented in the book does not claim to be the absolute truth, and is a small element of science in the area of theory of strength – soil mechanics. The basic meaning of submission is directed to the subjective judgment of the authors as to the nature of soil strength. The concept of the strength of the soil is characterized as a process of cumulative impacts of critical external and internal forces of gravity, in which at one or more sites progressive accumulation of plastic deformations occurs. Analytically, this situation in the ground can be described as the dependence of the critical external and gravitational forces, using strength parameters which include the angle of internal friction and specific cohesion. It is believed that the marketplace shears always rejected by an angle equal to $\theta = \varphi^*$ and $\theta = \left(\pi / 2 - \varphi^*\right)$. The physical meaning of the parameter φ^* is described in Chapter 1.

The book contains practical examples of solutions to various geotechnical engineering problems using the proposed theory of strength of soils.

It should be noted that complex physical processes such as the strength and stability of the soil were initially described in the classical scientific sources of Coulomb (1773), William Rankine (1857), and L. Prandtl (1921). Thereafter, these fundamental theories were developed in the works of V. Sokolovsky [10], K. Terzaghi [14], S. Golushkevich [4], V. Berezantsev [1], D.H. Shields [19], and others.

In drawing up the basic concept of the theory of strength of soils, the authors have reviewed the scientific papers of these researchers: D. Taylor [13], M. Malyshev [9], G. Klein [6], A. Sychev [11], Y. Zaretsky [5], N. Tsytovich [18], N. Puzyrevsky [23], S. Ukhov [16], F. Tatsuoka [22], I.Towhata [15], R.F. Craig [7], M. Braja [3] and others.

In Uzbekistan, the following scientists are engaged with the problems of soil strength and stress-strain state under static loads and seismic loads: T.R. Rashidov, T.Sh. Shirinkulov, H.Z. Rasulov, K.S. Sultanov, A.A. Ishankhodjaev, G.H. Hozhmetov, I.U. Mazhidov, M. Mirsaidov, F.A. Ikramov, M.M. Honkeldiev, I.I. Usmonhodzhaev, M.M. Yakubov, et al.

The authors will gratefully accept any criticisms and suggestions related to the content of this book. We can be contacted via SamGASI, the geotechnical laboratory of "Geofundamentproject." Location: 7000047, Lolazor 70, Samarkand, Uzbekistan, e-mail: uzssmge@gmail.com, Web site: www.geotechnics.uz.

The authors express their gratitude, for support in the writing and publication of this book, to the staff of the Geotechnics and Roads Department of Samarkand Architecture and Construction Institute and Ulugbek and the geotechnical laboratory staff of the "Geofundamentproject."

1 Experimental and theoretical studies of soil strength

1.1. Overview

It is known that under the classical definition of the term, "material durability or strength" is understood as resistance to mechanical influences from external forces without disturbing the material's original structure and continuity. The tensile strength of the material is characterized by the critical stress at which it is destroyed. *Since the physical condition and composition of soil is characterized as disconnected or weakly connected congestion in the amount of crushed mineral particles, so, when applied to such a multiphase term of strength of natural materials, we need to understand this term as resistance movement.* Progressive shear or movement corresponding to the limit stress of the soil is possible only after the integral accumulation of microshears and completion of the bulk soil compaction.

Upon reaching the critical values of shear displacement, the process of compaction is completed, and the ground is gradually transformed into a stage of disintegration. At this stage, there is an intensive development of plastic deformations of the soil on the most dangerous sites. In the process of shear, the soil starts to get harden and in time of reaching the critical point the soil transfers into the stage of loosening.

The term "plastic flow of soil" refers to progressive shear, irreversible movement of soil particles on concrete surfaces under the limit stress. This occurs when the lateral pressure coefficient reaches its minimum value, and the lateral expansion coefficient its maximum value.

For sandy soil in plastic flow, the transverse expansion coefficient reaches its maximum value, and for this reason, the volume extruded by the soil maximal normal stress is moved at an angle towards the smallest normal stress. Experimental methods for determining the direction of the displacement are described below (Figure 1.1).

For structured clay soils the formation of shear surfaces occurs after the destruction of the original structural links between the particles and their

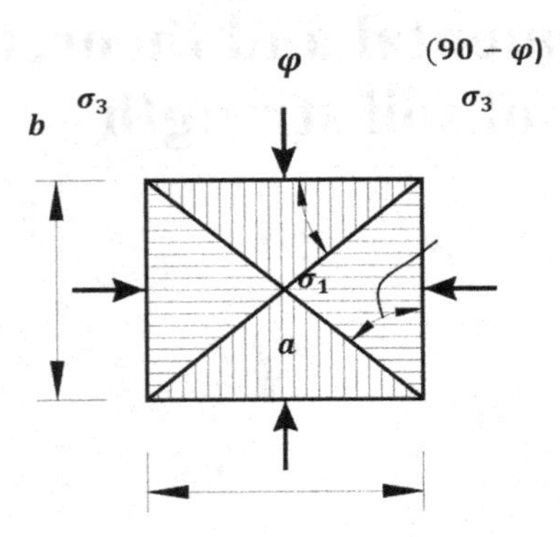

Figure 1.1 Scheme of the stress state and platforms inclined from the horizontal at an angle $\theta = \varphi$ inclined from the horizontal.

aggregates. Further accumulation of plastic deformations is accompanied by some compaction of the soil in the same way as in a granular soil formation. However, upon reaching its critical stress value, in contrast to sandy soil, the loosening process is practically not observed. Obviously, changes to these soils will happen after the destruction of the structural links and after the complete formation of the shear surface. The shear surfaces for structured clay soils do not extend over the entire volume as for sandy soils, it extends in places that come into contact with the weakest areas. That is why when testing of soils goes under the crushing conditions, inclined cracks appear on the surface of cylindrical samples. The process of destruction of moistened highly plastic clay soils takes place according to the laws of viscous plastic deformation.

Thus, all kinds of soil failure can be divided into hard (brittle), viscous and plastic. Hard (brittle) fracture types are predominantly non-moistened clay soils. In this case, the destruction of soil is due to the loss of structural integrity in some areas in the form of a soil (rock) that is cleaved into separate pieces. Therefore, the strength of such soils mainly depends on their cohesion. The nature of the destruction of granular and moist clay soils, including sand, belongs to the category of plastic and viscous-plastic. In this case, the strength of the soil is characterized by parameters such as the angle of internal friction, cohesion, coefficient of lateral pressure and maximum deflection angle of shear pads. For highly plastic clay soils of high humidity, the strength of soils is mainly characterized by viscous plastic properties. They appear as a process of long-term deformation at a constant pressure. The strength of such soils depends on parameters such as the angle of internal friction, cohesion and viscosity of the specific primer.

Some people believe that these basic properties of soils are a minor issue that can be neglected. In this case, the calculation of the stress-strain state of soils can be roughly considered as a continuous unbroken body. Naturally, this will simplify the tasks of assessing strength and deformability of soils. However, as will be shown below, this simplified approach distorts the processes in soils and leads to large discrepancies between the calculated and the actual results in practice.

For soils, especially granular ones, the term "voltage at the point" is arbitrary, since efforts in the array are distributed contacts of the mineral particles. Therefore, when we think about the state of stress in the elementary area, we mean some average state effort concentrated at this site (Figure 2.1). Just so, the nature of stress distribution in the soil body is different from the state of stress in unbroken array environments. However, it is clear that for each type of soil we cannot create separate functional dependencies, allowing us to estimate the stress state of the soil. But at the same time, at the decision of the individual in charge of geotechnical problems, we just cannot afford to ignore this important factor, for example, when calculating the stability of the body of hydraulic structures, or loaded with mounds of soil. Therefore, in all geotechnical research laboratories around the world, intensive research now aims at the development of the basic laws of soil mechanics to take into account the specific properties of soils.

As shown below, there are also grounds for the concept of sustainability. It is known that the resistance to continuous media is characterized by such a situation, when the body under the influence of external loads is in equilibrium, while maintaining a stable original form. For example, a mineral particle of the soil on the surface of the slope is in an extremely stable state, when the same particle is on the site deviated from the horizontal by an angle $\theta < \varphi$, and its position is considered stable. *The stress state corresponding to the beginning of a violation of the equilibrium state, in which there is a change of the original shape due to changes in the array of soil, is called critical.* For example, a block of soil limited by a natural slope, a soil bed with a cross section in the form of a trapezoid or a truncated cone with slope deviation angles $\theta < \varphi$ are stable forms for cohesionless soils (Fig. 1.26, 2.2). In contrast, for cohesive soils, there is no such restriction. However, in both cases, the loss of stability may occur at critical stress or under external stress. If the strength of soils can be described using the basic parameters of the model of Coulomb, i.e. the angle of internal friction and adhesion, then the stability, in our opinion, is an additional parameter of the lateral pressure coefficient of the soil, corresponding to the limiting stress state.

For soils as the dispersed environment, the coefficients of lateral pressure and the expansion are connected with the movement of soil particles in areas of potential shear surface. For this reason, both parameters are closely related to the magnitude of the angle of internal friction of the soil. In classical mechanics, the term "coefficient of lateral pressure" means the ratio of the increments of the smallest and the largest principal stress in the absence of lateral expansion. Primers, unlike the continuum mechanics in the limit state, go after the large plastic displacements and the formation of solid sliding surfaces. Therefore, the value of the coefficient of the lateral pressure, as will be shown below, largely depends on the value of the displacement and the deviation angles of the shear surfaces. Thus, for the soil, when solving geotechnical applications, you also need to know the ratio of the principal stress limit at which there is a loss of strength and stability. For example, the value of the coefficient of lateral pressure for fine sand, in the absence of lateral movements (the result of triaxial tests), is 0.17–0.2, while the value of the same coefficient in the limit state of stress is much smaller. According to the results of experimental studies, it was also found that, the deviation angles of the shear surfaces under active and passive loads diverge by $(\pi/2-2\varphi)$.

Our experiments have shown that the amounts of the coefficient of lateral pressure, which is equivalent to the limit of the active and passive loading, are different. A special case is the ratio of the limit values of the principal stresses in the inclined position of the formation. In this position of the sites, the influence of the ground friction force on the lateral pressure coefficient will be maximum and therefore the shear surface tends to take a position close to the horizontal direction.

Below, we regard the influence of the stress state on the strength and deformability of the soil. From the theory of classical mechanics it is known that the strained state in an elementary volume of soil can be conditionally divided into volumetric (normal) and shear (Fig. 1.2). In this case, shear strength contributes only to deformations of the soil shape and does not affect its volumetric component. This separation of stresses for soils is of fundamental importance, since the first is only volumetric, and the second, in contrast to continuous media, is both shear and volumetric.

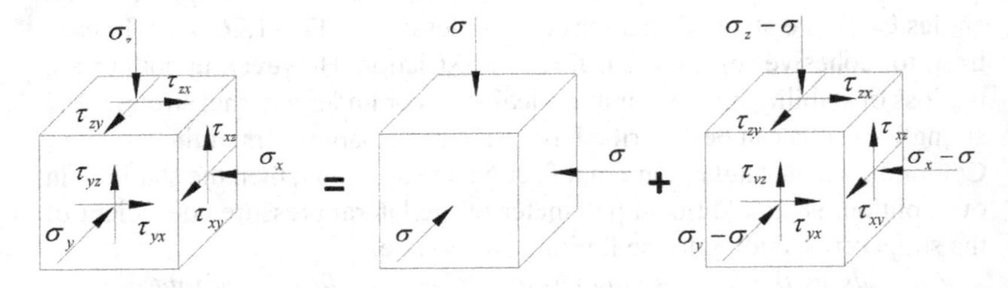

Figure 1.2 Decomposition of the stress on the volumetric and deviatoric tensor (shape change).

Volumetric soil deformation contributes to its compaction and hardening, whereas in the second part, divider stresses contribute to the accumulation of shear displacements accompanied by the same volumetric deformations. Moreover, in conditions of the maximum stressed state for loose soils, the soil's loosening is possible, increasing its volume.

When a volumetric stress state is applied to the ground, inside it the process of the emergence of microshears begins. If the elementary volume of the soil is affected only by hydrostatic stress, then the vector of the direction of movement of the particles tends to accumulate in the direction of a single point located in the center of gravity of this volume. In this case, all directions of the shear sites intersect at the center of gravity of the elementary volume. For example, if we take a ball as an elementary volume of soil, then the shear surfaces have the shape of a cone with a ball base. If a rectangular prism is represented as an elementary volume, then in this case the shear surfaces have the shape of a pyramid. All four pyramids with a pinnacle angle are connected at its center. Thus, an important conclusion can be drawn that for soil massifs with a rigid skeleton (coarse-grained sands, ground gravel) and for water-saturated clay soils, shear deformations prevail over volumetric deformations, and for porous low-moist clay soils, volumetric deformations prevail over shear deformations.

For plane strain as an elementary volume, we take a single thick rectangle. The direction of movement of elementary particles takes place on the surfaces of these areas. Shear surface in this case has inclined surfaces, and the deflection angle can vary depending on the orientation of the greatest principal stress (line 1 in Figure 1.3).

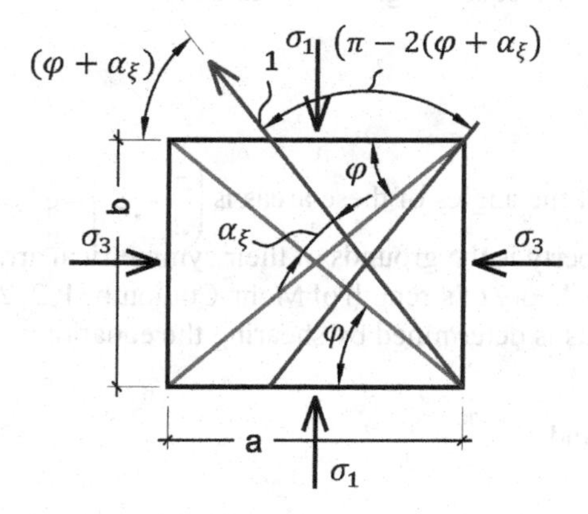

Figure 1.3 The circuit state of stress and the shear surface in the elementary volume of soil with active loading. 1 – the direction of movement of soil particles.

Given the accumulation of mineral particles in a confined space, their movement by increasing bulk power is limited. In this case, in the whole space of the elementary soil, volume quantities greatly exceed normal stresses, shearing their components. In this case, the functional dependence of soil strain on increasing hydrostatic stress tends to decrease.

Now imagine that these elementary volumes of soils are also affected by deviator voltages. In this case, at the initial stage of formation of shear surfaces, there is also an accumulation of volumetric deformations (compaction deformations); as shear stresses approach critical levels, and shear displacements accumulate, the growth of volumetric deformations stabilizes, and then the soil gradually moves into a stage of loosening (increasing in volume). At this stage, continuous shear surfaces are formed (line 1 in Fig. 1.3) and the soil goes into a plastic flow condition. The process of forming shear surfaces under extreme strengthen conditions will be shown below. As shown by experiments for soils, the angle of deviation of the shear surface and the movement of particles is somewhat smeared. Therefore, as with passive soil resistance, this angle is deflected relative to the site on which the maximum principal strengths act,

$$\theta = \varphi^* \tag{1.1}$$

where φ^* is the marginal deviation angle of the shear sites. The physical meaning of this parameter will be considered below. For example, when solving tasks of determining soil pressure on the fence, the deflection angles of these sites respectively with active and passive loads for mutually perpendicular platforms are equal (Fig. 1.3, 2.7 and 2.8):

$$\varphi^*; \left(\frac{\pi}{2} - \varphi^*\right) \tag{1.2}$$

The difference of the angles of these areas is $\left(\frac{\pi}{2} - \varphi^*\right) - \varphi^* = \frac{\pi}{2} - 2\varphi^*$.

The main property is the grounds of their symmetrical arrangement. Traditionally, on the theory of strength of Mohr-Coulomb [1, 2, 20], the orientation of these areas is determined by shearing the equation:

$$\beta = \frac{\pi}{4} + \frac{\varphi}{2} \text{ and } \alpha = \frac{\pi}{4} - \frac{\varphi}{2} \tag{1.3}$$

From this, it follows that in both the first and second cases, the deflection angle of shear areas does not coincide with the angle of internal friction of the soil. Analyzing the equations (1.1) and (1.2), we can conclude that in both cases the deviation angles platforms shear coincide only on the condition that:

$$\varphi^* = \theta^* = 30° \text{ and } \alpha = \frac{\pi}{4} - \frac{\varphi}{2} = 45 - \frac{30}{2} = 30°$$

$$\varphi^* = \theta^* = \frac{\pi}{2} - 30° = 60° \text{ and } \alpha = \frac{\pi}{4} + \frac{\varphi}{2} = 45 + \frac{30}{2} = 60°$$

In order to clarify these circumstances, the authors have conducted several experiments aimed at finding the causes of the mismatch between the theory of the strength of soils corresponding to the Mohr-Coulomb model and the actual experimental results. Below are the results of these experiments.

1.2. Experimental study of strength of soils using laboratory instruments

In order to establish the invariance of the strength parameters of soil in relation to different theories of strength under complex stress state, we carried out comprehensive tests of soils under: shear in one surface; in a triaxial pressure; torsion tubular samples in terms of volume compression; earth resistance under passive and active loading. Comprehensive tests were applied to fine sands of proluvial origin, related to the foothill area of the Samarkand plateau. The test soils were exposed to dry air and humid conditions. The density of the sand in all tests was assumed to be 1.48 g/cm^3. Below are the results of these tests.

Experiments 1 and 2

In the free rash of fine sand to the surface range, we have found that the angle of repose of the soil is (35°–36°), and its density is not more than 1.40 g/cm^3. The experiments with the sand on the devices of one-surface cut showed that the angle of internal friction is $\varphi = 39°$ (Figure 1.4, 1.26).

Experiment 3. Determination of the results of soil tests in a triaxial compression stabilometry type A

The type A stabilometer is designed to crush cylindrical soil samples. The design of this type of stabilometer allows you to mold sand of a given density in an air-dry state directly inside the rubber shell. The design soil scheme for triaxial tests is shown in Figure 1.6. Below we present the results of soil testing on this device.

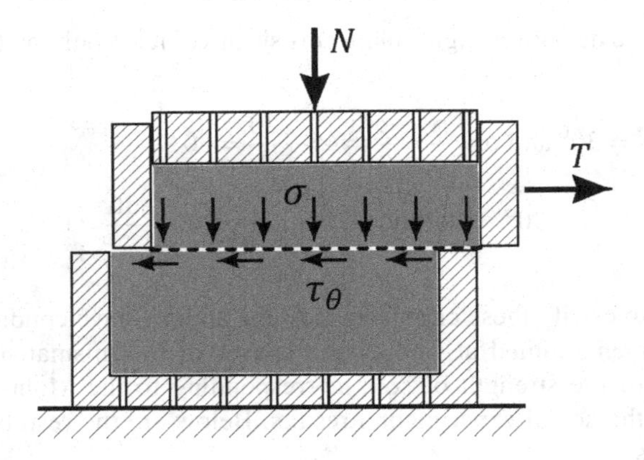

Figure 1.4 Design scheme of one surface shear.

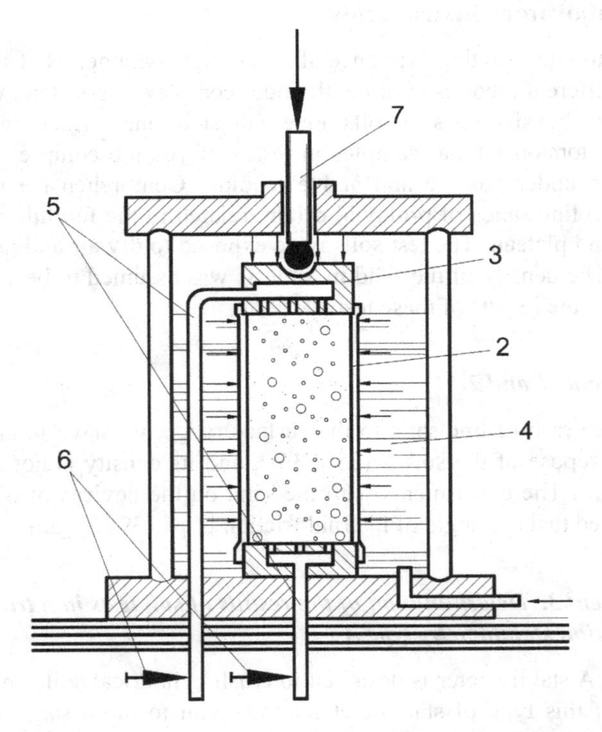

Figure 1.5 Scheme of stabilometer type A. 1-sample; 2-rubber sheath; 3-upper and lower pistons; 4-camera; 5-copper tubes; 6-taps; 7-stock.

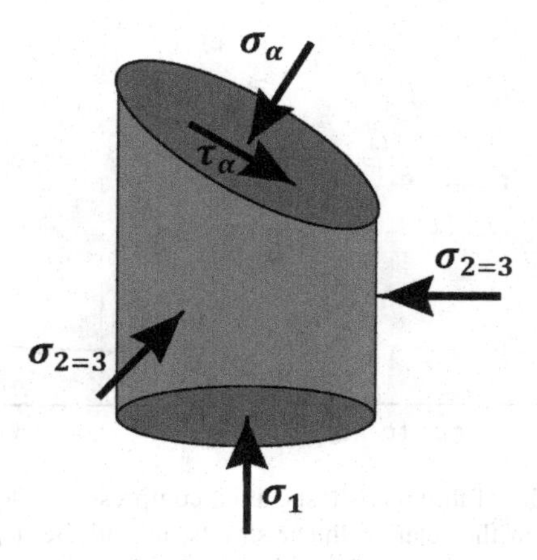

Figure 1.6 Design scheme of soil under triaxial tests. Axial compression $\sigma_{(1)} \geq \sigma_{(2=3)}$.

On these devices, a series of triaxial tests was carried out according to the scheme: crushing of cylindrical soil samples $\sigma_1 \geq \sigma_{2=3}$. The tests were performed on the same sand in an air-dry state, with an initial density of 1.48 g/cm^3. Geometric dimensions of the soil sample: diameter of the soil sample D = 35 mm, height H = 70 mm. The experiments were carried out according to the axial crushing scheme of the sample while maintaining the lateral pressures, i.e. $\sigma_1 \geq \sigma_{2=3}$. The tests were carried out at lateral pressures equal to $\sigma_{2=3} = 50;100$ и 200 kPa. The results of triaxial tests are given in table 1.3.

The results of these tests showed that, for example, at $\sigma_{2=3} = 200$ kPa and 500 kPa $< \sigma_1 \leq 760$ kPa, the soil sample gradually changed from a cylinder shape to a barrel shape. Moreover, it was not possible to establish clear traces of the surfaces of the shear. Such effects during triaxial testing of soils have been noted by many authors [5, 6]. Obviously, this is due to the conditions of soil testing. The value of the angle of internal friction according to the results of these tests, calculated in accordance with the Coulomb-Mohr theory of strength, is $\varphi = 35°$.

We should emphasize that in these tests, it is important to assess properly the criteria for the destruction of the soil. Various graphs are presented below for its evaluation. In particular, as the graph in Figure 1.7 shows, the dependence of the vertical deformation of the greatest principal stress takes the form of a power function. When the value of deviatoric stresses reaches (0.8–0.9) σ_u by limiting, the growth of volume deformation is stopped, and the ground is gradually transformed into a stage of disintegration (Figure 1.7).

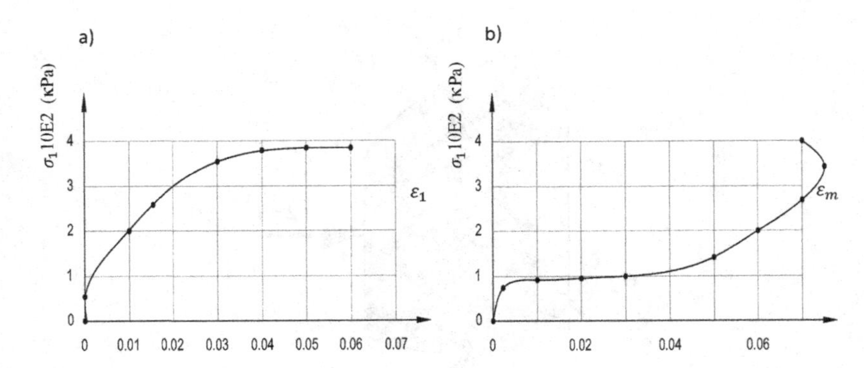

Figure 1.7 The results of three-axis tests, axial compression. Dependence of linear stresses on the relative linear strains (a) and the dependence of linear stresses and volume deformations under deviator loading (b).

We should note that stabilometric tests can accurately monitor the vertical and volumetric strain. At the same time, the lateral shear soil samples are determined by calculation. Due to the compressibility of the liquid in the chamber of the device and the uneven lateral movements of the height of the soil sample, it does not seem possible to determine the exact lateral movement in stabilometry. Taking into consideration the very low values of the displacement of the side faces, especially in the initial stage of deviatory loading, accurate measuring of the coefficient of lateral earth pressure and expansion is rather difficult.

For this reason, to determine directly the values of lateral shears and the coefficient of lateral expansion, we developed an experimental device based on uniaxial compression of a soil sample compressed by hydrostatic forces under negative atmospheric pressure using vacuum. Such tests require the development of special instruments and methods of preparation of soil samples for the test. A general view of the instrument and its circuit is shown in Figure 1.8. The device for the triaxial tests using vacuum consists of a body and a stand that allows vertical axial loading of the soil sample. The device itself consists of a lower housing clamped therein with a rubber skin (1). The rubber sheath is pulled over the bottom of a metal ring (2) and by means of rubber gaskets (3) is clamped with nuts (4) to the housing (1). The soil is placed inside the rubber shell on a perforated stamp (5), and on top of the one-piece die (disk) (6) with side slits for tightening the clamp rubber. The lower perforated die (5) rests on the housing, the nozzle (7) is connected to a vacuum unit using a reinforced hose. Soil is formed inside the metal tube (8). Soil is put inside a metal pipe (8). To do this, the metal pipe (8) with the lower end is first installed on the

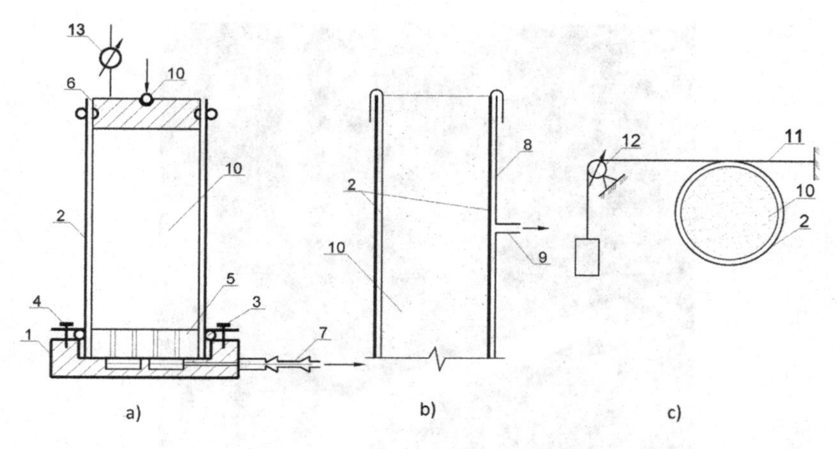

Figure 1.8 The triaxial test scheme, σ_m, is created with the help of vacuum (a); scheme for filling the soil in the chamber (b); and a scheme for measuring lateral movements with the help of Maximov's deflectometer (c).

device, so the rubber shell must remain inside the pipe and stretch onto it using the nozzle (9), installed on the side surface of the pipe. Air is sucked out until the rubber shell is pressed against the inner surface of the pipe. Further, sand with a given density is poured into the pipe and a stamp is installed on its upper end (6). After that, the metal pipe is removed by moving up. To center the axial load, a metal ball (10) is installed on top of the stamp and the device body with soil is installed to the center of the loading device.

Vertical deformation of the soil sample is measured by using a measuring device (13) fixed to the upper end of the boot device. Axial movements of the soil are determined with an accuracy of 0.01 mm. The lateral displacement is measured by means of a deflectometer and tensioned steel filament in a loop passing through the side surface of the sample (11). The measuring belt is installed in the center of the height of the soil sample. One side of the belt is fixed to an unmoved rod attached to the body, and the other is freely suspended from the pulley of the deflection meter (12) attached to the tripod. The accuracy of the division of the Maximov deflection meter is 0.1 mm. The accuracy of the division of the vacuum manometer is 0.5 kPa. The experimental device is shown in Figure 1.9.

Axial compression tests performed on soil were analogously designed for triaxial tests. First, hydrostatic volumetric strength was transferred to the soil sample to cause deformation. After stabilizing (completing) of the soil deformations process, axial loads were done to the soil sample, with pressures which are not exceeding 5–10 kPa.

Figure 1.9 Type of device designed to test triaxial soil samples, compressed by body forces with the help of a vacuum.

We have tested soil samples with a diameter of 70 mm and a height of 110 mm. The initial density of fine sand stabilometric take similar tests, i.e. 1.48 g/cm^3.

Lateral displacement and deformation of the ground were determined by the equation:

$$\Delta d_x = l_x / \pi \tag{1.4}$$

$$\varepsilon_x = \frac{\Delta d_x}{d} \tag{1.5}$$

where l_x – a move in the tangential direction (index of deflectometer).

The results of the experiments

The experiments were carried out under hydrostatic compression of the soil sample $\sigma_m = 50;90$ kPa. According to the results of triaxial compression $\sigma_1 > \sigma_{2=3}$, it was found that linear movements have a different pattern. In particular, it was found that the lateral displacement at the initial axial loads

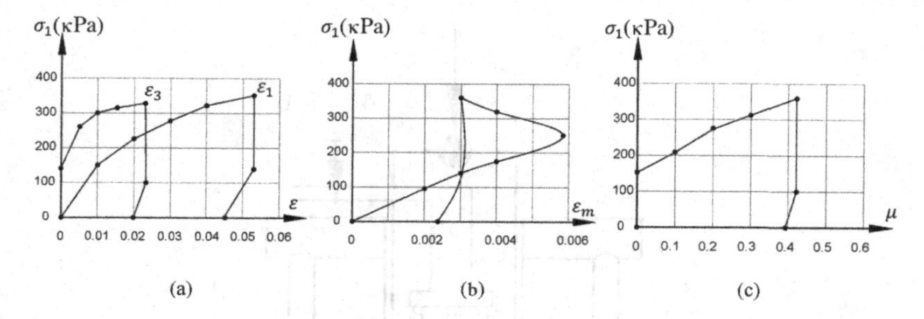

Figure 1.10 The results of three-axis tests. Dependence of the axial stresses on the relative axial displacement (a), volume deformations (b) and the graph of the change in the coefficient of transverse deformation (c) under deviatoric loading.

Table 1.1 The test results of sandy soils for triaxial compression

Volumetric voltage, σ_m (kPa)	50	90
Limiting axial stresses, σ_1 (kPa)	190	329
The angle of internal friction of Coulomb Mohr φ_M	36	36
Axial stress related $\varepsilon_{2=3} \approx 0$	90	140
The averaged value of the coefficient of expansion of the cross, μ	0.14	0.28
The averaged value of the coefficient of lateral earth pressure under the limit state of stress	0.26	0.27

does not develop as rapidly as axial displacement. This situation corresponds to the condition of compression when $\sigma_1 > \sigma_{2=3}$, $\varepsilon_{2=3} \approx 0$. With further increase in the axial stresses, side strain begins to appear, and when one approaches the point of destruction of the ground, it begins to grow rapidly. From this point on, the intensity of the lateral deformation is greater than that of the vertical deformation. For this reason, when approaching the critical state of stress, the soil will expand and disintegrate. Graphs of the change in volume and linear deformation during deviatoric loading soil are shown in Figure 1.10. Table 1.1 presents the main design parameters of ground deformation. In particular, modules linear cutting deformation corresponding to an axial stress of $\sigma_1 = 300$, respectively: $E_1/E_3 = 300/75$. Value limiting stresses in these trials correspond to $\sigma_3/\sigma_1 = 90/320$ kPa.

Experiment 4

Determination of soil strength and lateral pressure coefficient according to the test results under conditions of triaxial compression in a type B stabilometer (Fig. 1.11).

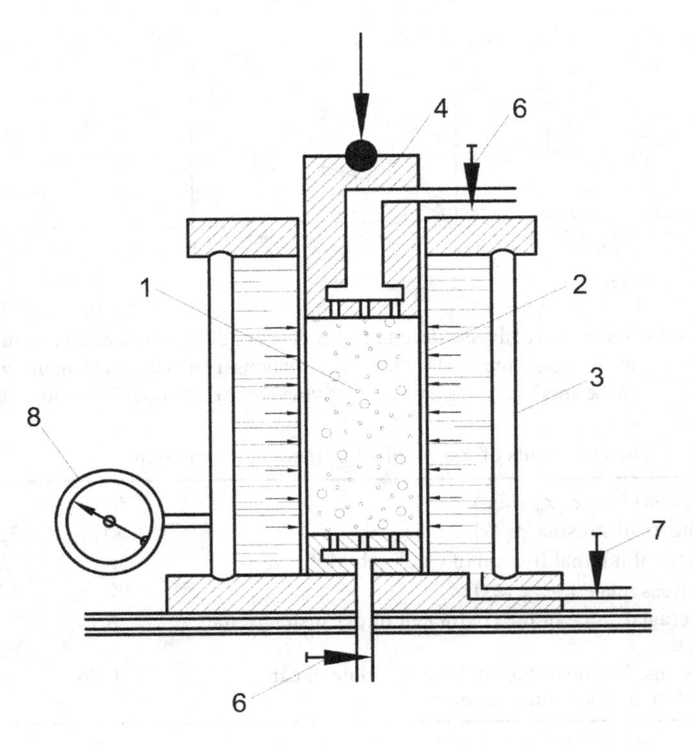

Figure 1.11 Driving triaxial type B. 1-design sample; 2-rubber sheath; 3-upper and lower pistons; 4-movable piston; 6 and 7-taps; 8-gauge.

This type of stabilometer, in contrast to the above, allows you to independently adjust lateral stresses and measure under compression conditions (with limited lateral movements) lateral stresses of the soil. For a soil sample of a given density, the sands were moistened and molded in a metal pipe, the inner size of which coincides with the size of the soil sample. A soil sample of a given shape was obtained after freezing it at a negative temperature. The soil after installing the stabilometer in the instrument chamber was first thawed and then tested.

The geometric dimensions of the soil sample were: soil sample diameter D = 70 mm, height H = 140 mm. Experiments were carried out according to the scheme of axial compression (crushing) in the absence and presence of the conditions of lateral movement, and also provided for stretching of the soil sample. The tensile test was carried out as follows. First, the soil sample was subjected to a comprehensive volume compression, $\sigma_m = 200$ kPa, then a gradual decrease in the vertical thrust to $\Delta\sigma_{2=3} = 104$ kPa.

The lateral ground pressure remained unchanged and equal to $\sigma_{2=3} = 200$ kPa. The destruction of the soil sample was achieved by stepwise increasing of the lateral pressure stages to $\Delta\sigma_{2=3} = 40$ kPa.

The results of these experiments are shown in Table 1.3. The results showed that the ratio of the limiting voltages (passive lateral pressures $\sigma_1 < \sigma_{2=3}$) corresponds to $\sigma_1 / \sigma_{2=3} = 104/460$ kPa. The destruction, in contrast to the testing scheme crush, occurred quickly. The angle of internal friction in the results of these tests, calculated in accordance with the theory of Mohr-Coulomb strength, is equal to $\varphi = 39°$, and the ratio $\sigma_1 / \sigma_{2=3}$ in a state of stress is equal to the limit of 0.23.

Experiments to determine the coefficient of lateral pressure in the absence of lateral movement were conducted using the compression mode, with the compression chamber in a closed system. The soil sample was pre-crimped by a hydrostatic pressure of 25 kPa. Next, using a crane, a hydraulic capacity triaxial disconnected from a volumeter otherwise created conditions limiting the total lateral movements $\varepsilon_{2=3} = 0$. Lateral earth pressure was measured by a gauge fixed to the housing triaxial. The test results showed that by limiting the lateral movement of soil (condition of compression) at a vertical pressure of $\sigma \leq 250$ kPa, the lateral pressure coefficient was $\xi = 0.12$. When $250 \leq \sigma \leq 600$ kPa, this figure increased to a value of $\xi = 0.17$. Schedule changes in the coefficient of lateral pressure from the linear strain under triaxial compression and tension for the tested sand is shown in Figure 1.12. In Figure 1.13 we show a graph of the lateral earth pressure during deviatoric loading.

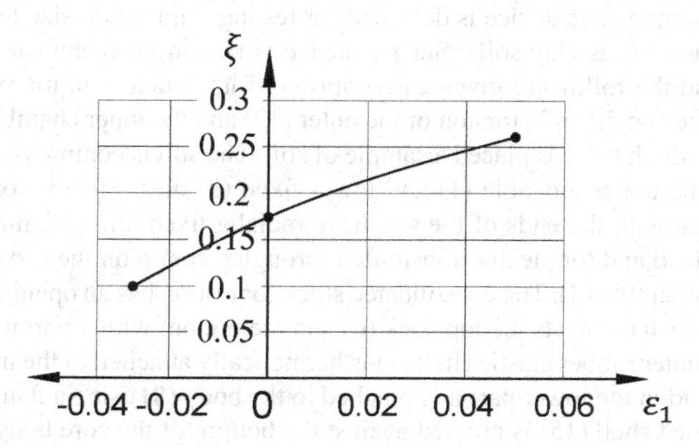

Figure 1.12 Schedule changes in the coefficient of lateral pressure from the linear strain under triaxial compression and tension.

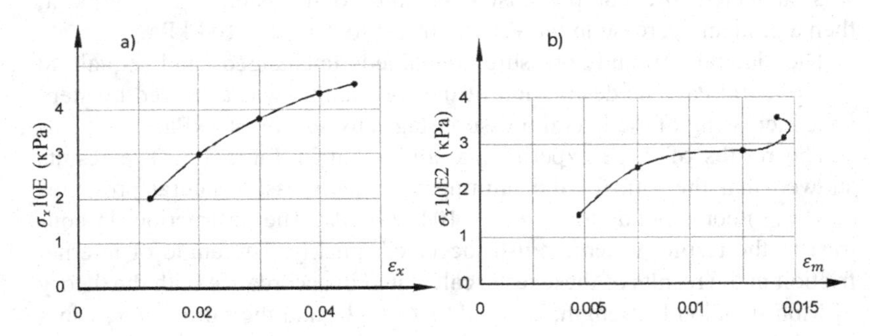

Figure 1.13 The results of three-axis tests. Dependence of linear stresses σ_{hot} on relative linear deformations ε_x with lateral compression of the soil sample (a); dependence of relative volume deformations ε_m. on deviatoric stresses during lateral compression of the soil sample (b).

Experiment 5. Determination of soil in a clean shear on the instrument triaxial torsion

For comparison of the results obtained, we also conducted three-axis test with torsion on a soil sample in the shape of a thick-walled tube. The dimensions of the soil sample are: outer diameter D = 90 mm, inside diameter d = 50 mm, height H of the hollow cylinder = 90 mm and wall thickness 20 mm (Figure 1.14). The Moscow Scientific Research Institute "Hydroproject," named after Gjuk (Russia), has developed this instrument design. Applied to tubular samples of soil, the device allows one to adjust independently the internal, external and axial pressure and create shear stresses at the ends of the soil sample. The device is designed for testing sand (grain size less than 1 mm), as well as clay soils. Such a device is rare in geotechnical laboratories, and the following gives a description of its structure in more detail. The device consists of a torsion of the outer (12) and the inner chamber (11), between which there is placed a sample of soil. The soil is compressed at the ends of the upper movable (1) and lower fixed (2) dies. Stamps for better connection with the ends of the soil have radially fixed knives 2 mm high. Vertical load and torque are transmitted through a stamp on the rod (3), the plunger assembly (4). The consolidated stock for testing has an opening (5) in height, which connects the top cock (6) and the bottom with a porous stamp (1). The outer rubber elastic shell (7) is hermetically attached to the movable die (1), and in the lower part it is pinched to the body (8) using a flange. The inner closed shell (15) is pressed against the bottom of the core body (9) by means of a clamping nut (10). To prevent torsion of the core when tightening, the clamping nut is installed between a set of metal balls. The loading

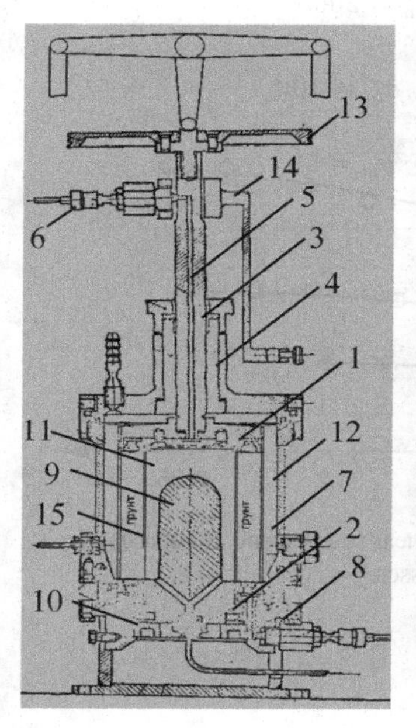

Figure 1.14 Scheme and general view of the triaxial torsion compression device with torsion for testing soil samples for shear.

of the sample is carried out hydraulically and mechanically. The fluid under pressure in the inner (11) and outer (12) the cameras of the device come from volumeters consisting of measuring glass tubes. The volumeter is connected to the receiver with the compressed air on one side and with the oven on the other. The air pressure in the receiver is controlled using a manometer. The torque is generated by applying horizontal forces on the pulley (13).

An LED with an accuracy of 0.01 mm for fixing the axial strain is fixed in the nozzle (14). The angular movement of the pulley is determined by the deflectometer, with divisions of 0.01 mm.

For manufacturing the soil sample in the form of a thick-walled tube, special forms and supplies are required. For testing fine sand, it is pre-moistened to the optimum moisture content. Further, the sand layers are fitted into a mold and compacted under the same load. After the completion of molding, the ground was frozen at subzero temperatures. Next, the frozen soil sample was taken from the mold, and the device was installed in the camera (Figure 1.17). The device was installed on the ground and the base frame, and by means of copper tubing was connected to the volumeter (Figure 1.14).

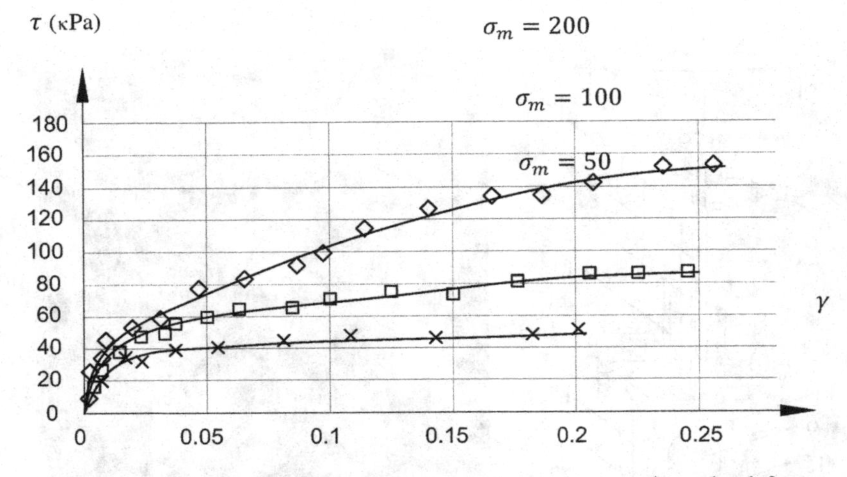

Figure 1.15 A plot of the relationship between shear stresses and angular deformations for different mean normal stresses.

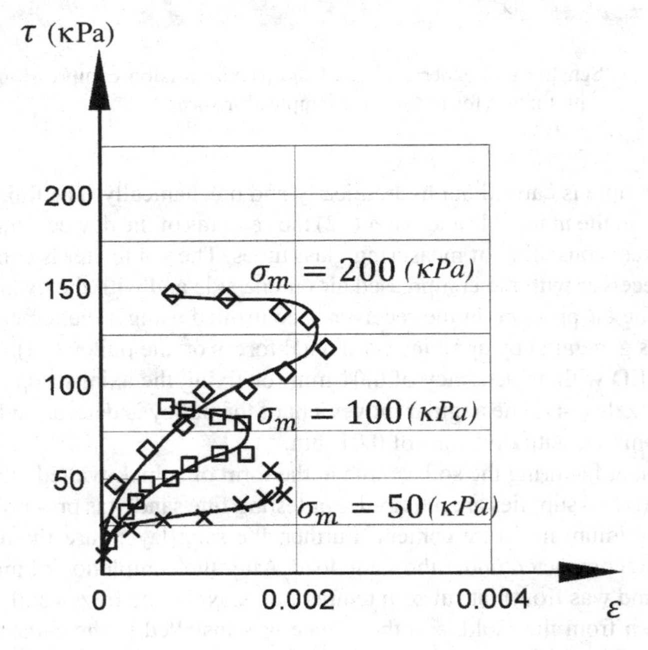

Figure 1.16 Graph of the dependence of linear deformations on the magnitude of angular deformations for various mean normal stresses.

Figure 1.17 General view of the soil sample prior to the test (a) and after the test (b).

To determine the resistance to torsion of the rubber membranes in the device, we installed in place of the soil sample rings of fluoroplast with minimal friction. The resulting dependence of the resistance to torsion on the shell's twist angle was taken into account when analyzing the results of the experiment. Testing began after a full thawing of the soil. Changes in the geometric dimensions of the soil sample were measured with the following instruments.

Changes in the external diameter of the soil sample:

$$\frac{\pi}{4}\left(D_{2a}^{2}-D_{1a}^{2}\right)\left(H-S\right)=S*A_{wm}+V_{a} \tag{1.6}$$

$$D_{2a}^{A}=\sqrt{D_{1a}^{2}\mp\frac{4}{\pi}\frac{\left(S*A_{wm}+V_{a}\right)}{\left(H-S\right)}} \tag{1.7}$$

Changes in the internal diameter of the soil sample:

$$\frac{\pi}{4}\left(d_{2i}^{2}-d_{1i}^{2}\right)\left(H-S\right)=V_{i}$$

$$d_{2i}=\sqrt{d_{1i}^{2}\mp\frac{4}{\pi}\frac{V_{i}}{\left(H-S\right)}} \tag{1.8}$$

where $D_{1a}^2 = 81 cm^2$; $d_{1i}^2 = 25 cm^2$; H = 9 cm, height of the soil sample; Vertical S-precipitate (displacement of the rod 11); $A_{wm} = \frac{\pi D_{wm}^2}{4}$ – Area of the stem (11); V_a, V_i – fluid flow in the (outer and inner chambers) volumetry.

The state of stress for the shape of the sample in the form of a thick-walled pipe is conveniently represented in cylindrical coordinates.

In our experiments, tests were carried out only at hydrostatic pressures in the camera of the device, following the condition $\sigma_r = \sigma_\theta = \sigma_z$. The shear stresses acting on the ends of the soil sample were determined by the equation:

$$\tau = \frac{4}{\pi} \frac{R_{wk} * T}{r_m \left(D^2 - d^2\right)} = \frac{4}{\pi} \frac{M}{r_m \left(D^2 - d^2\right)} \tag{1.9}$$

where $R_{wk} = 78\, mm$, the radius of the pulley (13); T – load on the suspension; $r_m = 35\, mm$, the average radius of the soil sample; M – external torque load exerted on the pulley (13).

According to the measurement of geometric forms, we define all the types of deformation:

Volumetric deformation

$$\varepsilon_m = \frac{V_a + V_i + S * A_{wm}}{V} \tag{1.10}$$

Angular deformation

$$\gamma = \frac{2\pi}{360} \frac{r_m * \varphi}{(H - S)} \tag{1.11}$$

$$\varphi = \frac{360}{2\pi} \frac{L}{R_{wk}} \tag{1.12}$$

where L is angular displacement (indication deflectometer).

The results of these experiments showed that the angle of internal friction of soil is equal to the value $\varphi = (39 \div 40)^\circ$. If suspended, as a fixed deformation take $\gamma = 0.1$, the shear moduli for $1 - \sigma = 200$ kPa; 2 – σ = 100 kPa and 3 – 3 – σ = 50 kPa, respectively, are equal to G = 110/0.1 = 1.1; G = 74/0.1 = 0.74 mPa; G = 40/0.1 = 0.40mPa.

The functional relationship between the deflection angle of the most dangerous areas, constituting the stress τ = f (σ) and the main stress, will be discussed below in section 2.3 (Case 4).

When the torsional soil sample is first loaded into a tube, a sealing effect is observed. When the shear stresses approach the critical values, there is loosening (the effect of dilatancy). In our experiments, the extra bulk soil compaction in pure shear before its destruction does not exceed 0.002, or 0.2%. In all tests, in the approach to the ground's stage of destruction, there was a slight loosening (Figure 1.16). The effect of soil compaction under the action of shear stresses is associated with the emergence of movements of particles in parallel planes. At the same time, soil particles are compacted into a compact position. However, with critical loads, such sealing becomes impossible, and any slight increase in soil shear moves into a phase shear of undamped motion. In this, the first compaction process is completed, and the ground is gradually transformed into a stage of disintegration (Figure 1.16).

Thus, the shear displacement of particles (strain) of soil at a constant average and shear stresses depends on the initial density of the soil. Such dependence can be analytically expressed as follows:

$$\gamma = f(\tau, \sigma) \tag{1.13}$$

This simple functional dependence of the shear modulus of the medium voltage, in a linear calculation, can take in the following form:

$$G = (G_0 + G_1 * \sigma_m) \tag{1.14}$$

where G_0, G_1 – parameters of the equation. For unstructured soils, $G_0 \cong 0$. For fine sand tests, $G_1 \cong 6$. For example, $\sigma_m = 50;100;$ and 200 kPa. G = 300, 600 and 1200 kPa.

We suggest that volumetric soil deformation should be defined separately on two parameters: volume compression only on the hydrostatic stress and the effect of the deviator (tangential) stress:

$$\varepsilon = \frac{\sigma}{k} \mp \frac{\tau}{m};\ m = f\left(\frac{\rho_{max} - \rho}{\rho_{max} - \rho_{min}}\right) \tag{1.15}$$

For fine sands, the bulk modulus of compression is k = (2.5–3.5) mPa and the bulk deviator modulus is m = (5–6) mPa i.e. the condition m/k = (1.7 ÷ 2) is met.

Given that the unit torsion test of cylindrical samples of soil in geotechnical laboratories is maintained, the results of these experiments are of particular interest for expert laboratories, which rarely use these devices, unlike stabilometric triaxial tests of uniform stress state. Therefore, the results of these tests conducted by us on fine sand having a density of 1.48 g/cm³ are shown in Table 1.2.

Table 1.2 The results of tests on sandy soils for pure shear

The bulk voltage σ_m (kPa)	*Limiting the shear stresses τ_u (kPa)*	*Angle of internal friction*	*Conventional fixed angular deformation γ*	*The shear modulus (loading stage) G (mPa)*	*The shear modulus (unloading) G_{ynp} (mPa)*
50	46	42	0.1	0.5	2.1
50	46	42	0.1	0.5	2.05
50	45	42	0.13	0.5	2
100	83	40	0.11	0.65	2.5
100	83.5	40	0.12	0.73	2.8
100	85.5	40	0.16	0.8	3.2
100	86	41	0.2	0.75	3.1
200	160	39	0.23	1.16	5.33
200	165	40	0.21	1.2	5.1

1.3. Experimental studies of strength, stability of soil and location of surface changes

The issues of the influence of the stress state on the strength and on the main parameters of the soil strength equation were considered above. In this section we consider the influence of parameters of soil strength by a factor of lateral pressure on the deflection angle of the shear surfaces.

It is known that the strength properties largely influence the accuracy of the calculated values of maximum loads transmitted to the ground. Determining the relationship between the strength characteristics of the soil and the limiting of the load on the array of soil is a difficult task, because to assess properly the real stress state in the array is not possible. For this reason, the authors proposed to conduct such tests on a special stand, where the amount of soil on the sides of the glass wall is limited. In this case, the ground state of stress, in the form of elongated prisms loaded at the ends of the principal stresses, is assumed to be known. The laboratory tray consists of a hard metal casing bottom and two sidewalls of fixed glass. The cross section of an end of the tray measures 90 × 90 mm. The tray has a length of 220 mm. The tray allows experiments to determine the stability of the soil as a scheme of active and passive loading. During the test, movements of the two mutually perpendicular main areas were measured with dial gauges, with divisions of 0.01 mm. The load was transmitted to the ground by means of the loading device, designed for the standard shear test. The results of these experiments are shown below. The construction of such a device is shown in Figure 1.18.

Figure 1.18 The instrument for determining the passive resistance of soil.

Experiment 1

The purpose of experiment is determination of strength, soil stability, and position of shear surfaces at passive pressure.

To solve this problem, the internal volume of the tray was filled with soil sample. Grunt – fine sand in the air to a dry density of $\gamma = 1.48$ g/cm^3. In these experiments, the ground pressure was passed through a side of the square with a stamp size of 90 × 90 mm. It was considered as cases of load and non-load to the surface layer of soil.

Vertical force at the depth of *h* is equal to:

conducted with reference to experiments

$$N_1 = \frac{b^3\gamma}{tg\varphi} + q$$

in relation to the plane problem

$$N_1 = \frac{b^2\gamma}{tg\varphi} + q$$

where b – thickness; q – the value of load on the soil surface.

Let's define the estimated balance of forces acting on the surface *AC*:

$$\frac{\tau}{\sigma} = \frac{\Sigma H}{\Sigma V} = \frac{N_3 sin\theta - N_1 cos\theta}{N_1 sin\theta + N_3 cos\theta} = \frac{tg\theta - \xi_N}{\xi_N tg\theta + 1} \tag{1.19}$$

where $\xi_N = N_3 / N_1$ – coefficient characterizing the ratio of the horizontal and vertical forces; $\theta = \varphi^*$ – the angle of deflection site of shear.

The experiments showed that under the stress limit state, the surface movement of soil particles varies nonlinearly. This dependence can be described by the following functional relationship:

$$y = a * \ln x \tag{1.20}$$

If we take the coefficient characterizing the nonlinearity function $F_s = \varphi / \theta$, then

$$\frac{\tau}{\sigma} = \frac{tg\theta - \xi_N}{\xi_N tg\theta + 1} = \frac{tg\varphi}{F_s} \tag{1.21}$$

where φ is the angle of internal friction corresponding to Coulomb strength theory.

If in equation (1.21) we take as an unknown the angle of deviation of the site that is different from φ, for example θ, then

$$tg\theta = \frac{\xi_N + \dfrac{tg\varphi}{F_s}}{1 - \dfrac{tg\varphi}{F_s}\xi_N} \tag{1.22}$$

For example, if $\xi_N = \dfrac{12.6}{93.8} = 0.14$; $\varphi_k = 39°$ and $F_s = 0.95$, $tg\theta = 1.13$, $\theta = 49°$.

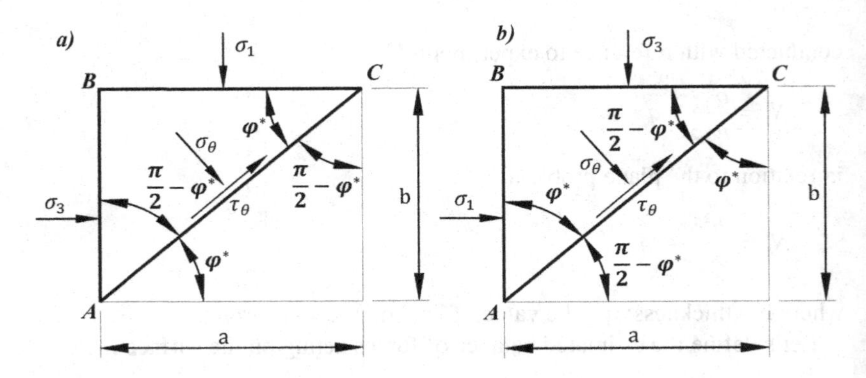

Figure 1.19 Surface position shears in the elementary volume of soil with the active (a) and passive (b) loadings corresponding to the limit stress state.

Then we solve this problem with the help of stress. Imagine that we know the ultimate state of stress at a point where there are major voltage $\sigma_{2=3}$, σ_1 the lateral pressure coefficient ξ. We also know the angle of deflection surface shear θ. We assume that the direction cosines of the normal drawn to the site are offset $l = \frac{\pi}{2} - \varphi$ and $s = \varphi$. In this case, we use well-known classical mechanics equations [12] and determine the components of the voltage σ_θ and τ_θ, acting on the ground with an inclination angle θ:

$$\tau_\theta^2 = \sigma_1^2 l^2 + \sigma_3^2 s^2 - \sigma_\theta^2 \tag{1.23}$$

$$\sigma_\theta = \sigma_1 l^2 + \sigma_3 s^2 \tag{1.24}$$

The component of the total stress vector on an inclined platform can be expressed in terms of the initial principal stresses.

$$P = \sqrt{\sigma_1^2 l^2 + \sigma_3^2 s^2} = \sqrt{\sigma_\theta^2 + \tau_\theta^2} \tag{1.25}$$

On the surface, the shear strength should be satisfied with Coulomb soil:

$$\tau_\theta = \sigma_\theta tg\varphi \tag{1.25}$$

The joint solution of equations (1.23–1.25) must satisfy the results obtained from the experiments, that is, Condition (1.25). Condition (1.25)

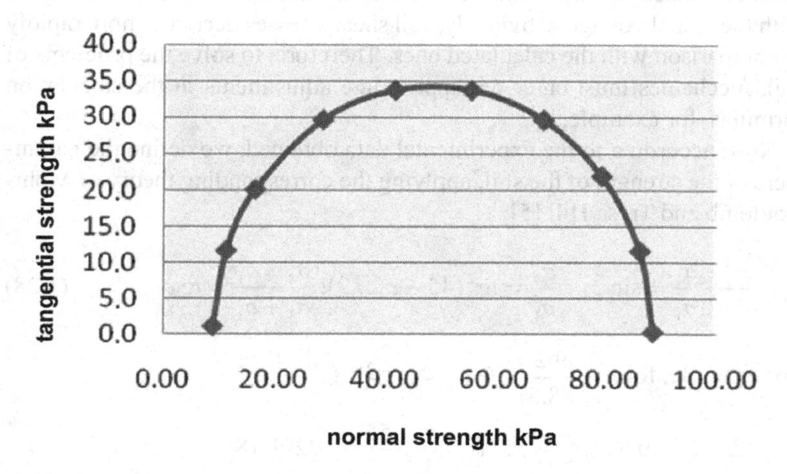

Figure 1.20 The diagram of the change in the stress state of the soil at a point.

must be carried out on site with direction cosines equal to $l=\frac{\pi}{2}-\varphi$ and $s=\varphi$. This is possible only when the shear stress intensity decreases by a factor of

$$k_\tau \cong 1-\xi \tag{1.26}$$

$$\tau_\theta = k_\tau\sqrt{\sigma_1^2 l^2+\sigma_3^2 s^2-\sigma_\theta^2}=\sigma_\theta tg\varphi=\left(\sigma_1 l^2+\sigma_3 s^2\right)tg\varphi \tag{1.27}$$

The law of change of the tangent and normal stresses in view of factor k_τ is presented in Figure 1.20.

Example

According to soil tests for passive pressure we have obtained $\sigma_1 = 9.2kPa$, $\sigma_3 = 88.6kPa\ \theta = 51°$. The direction cosines of the shear platform are equal to $l=\varphi^* = cos51° = 0.618$ и $s = cos\left(\frac{\pi}{2}-\varphi^*\right) = cos\left(\frac{\pi}{2}-51^*\right) = 0.79.$

On this platform $\tau_\theta = k_\tau\sqrt{\sigma_1^2 l^2+\sigma_3^2 s^2-\sigma_\theta^2} = 38kPa;\ \sigma_\theta = \sigma_1 l^2+\sigma_3 s^2 =$ $88.6*0.37+9.2*0.63 = 39kPa$, $tg\theta = \frac{\tau_\theta}{\sigma_\theta} = \frac{38}{39} = 0.97$. Compare $tg\theta = 0.97$ or $\theta = 44°8\ tg\varphi = 0.81$ и $tg\varphi = 0.81$, or $\varphi_k = 39^0\ k_\tau = \frac{0.97}{0.81} = 1.2.$

This example, like the previous one, shows that the state of stress at the site determined by the methods of classical mechanics does not coincide with the actual voltage. Obviously, soil shear stresses decrease more rapidly in comparison with the calculated ones. Therefore, to solve the problems of soil, mechanics must enter the appropriate adjustments in the calculation formulas, for example, k_τ.

Now, according to the experimental data obtained, we define the parameters of the strength of the soil, applying the corresponding theory of Mohr-Coulomb and Tresc Hill [5]:

$$\frac{\sigma_3-\sigma_1}{\sigma_3+\sigma_1} = \sin\varphi_M;\ \frac{\sigma_3}{\sigma_1} = tg^2(45-\varphi_M/2);\ \frac{\sigma_3-\sigma_1}{\sigma_3+\sigma_1} = tg\varphi_T \tag{1.28}$$

For example, for $\xi = \frac{9.2}{88.5} = 0.11;\ \varphi_k = 39°;$

$$\sin\varphi_M = 0.81;\ \varphi_M = 55°\ \theta = \frac{\pi}{4}\pm\frac{55°}{2} = 72° \text{ и } 18°.$$

$$tg\varphi_T = 0.81\varphi_T = 39°;\ \theta = 45°$$

The experimental results and the calculated values of the angles of internal friction, and the coefficient of lateral pressure, are shown in Table 2.1.

Based on the experiments, the following was established: the direction of the AC shear pads under active and passive loading has a deflection angle with respect to the horizon, respectively: $\theta = \varphi^*$ and $\theta = \frac{\pi}{2} - \varphi^* = \frac{\pi}{2} - (\varphi + \alpha_\xi)$. The direction of the AC shear pads under passive loading is rejected with respect to the pads where the largest principal stresses act at an angle $\theta = \varphi^*$ (Figure 2.7, 2.8).

Experiment 2

Experiment task is determination of strength, soil stability and position of shear surfaces at passive pressure for the case of an inclined formation located at an angle $\beta = \varphi$.

The main purpose of these tests was to determine the direction of the shear surface or ground movements. For this purpose we consider the limiting state of stress on the device, simulating an infinite slope with an angle of $0 < \beta \leq \varphi$. This problem was first considered in [15]. The same problems are considered and regarded in [2, 10, 11]. In all these works it is assumed that the maximum shear surface is inclined from the horizontal at an angle of $\beta = \theta = \varphi$. However, in these studies in the preparation of the equations of equilibrium resistance, the factor of layer caster efforts is not taken into account. In fact, we consider the continuous flow layer of soil on a slope with an inclination angle of $\theta > \varphi$.

The main distinctive feature of such an experiment compared to others is the imitation of the stressed state of the surface inclined layer of slopes. During the passive loading of an inclined bed of soil, any longitudinal movements of it are accompanied by the extrusion of sand in a direction perpendicular to the inclined surface. In this, sand pours down the slope. With critical passive loads, this process goes into the area of flow, and otherwise there is a complete loss of stability of the slope.

Given the importance of this for solving applied problems of geotechnical engineering, laboratory model tests were conducted for sand. For the deflection angle of the tray relative to the horizon, the angle of repose is assumed to be equal to 360. It is noteworthy that in this case the surface of the shear speaker has a curved shape. Approximation of this line can be roughly taken as a straight line parallel to the horizon line.

The directions of plastic displacements of the soil practically coincide with the directions of the shear surfaces. The scheme of plastic displacements of the soil along shear surfaces is shown in Fig.1.21. The experimental stand is presented on Fig. 1.22.

The experiment was conducted in the following manner: on the side of the ground by the end of the square stamp, in small steps a load $N_{3\beta}$ is

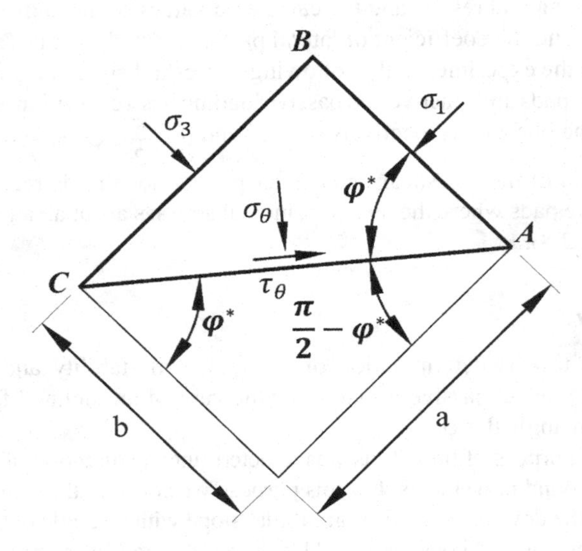

Figure 1.21 Scheme of passive resistance of soil under inclined position of the layer of soil.

Figure 1.22 Instrument for determining the passive soil pressure in an inclined position.

applied. In this experiment, the deviation angle from the horizon of the tray coincides with the angle of repose, i.e. $\beta = \varphi = 36°$. With a gradual increase in pressure on the end face of the soil sample, deformation accumulates in mutually perpendicular directions. The ratio of relative deformation during loading increased from 0.45 to 0.8. When approaching the critical stress value $\sigma_3 = 0.9 * \sigma_{3u}$, gradually the ground goes into plastic flow and enters an unstable state. Extruded sand begins to fall out of the tray. The angle of repose of the soil and its form remains unaltered for the entire length of the experiment. The surface changes in the conditions of the limit state of stress tend to assume a position close to the horizon.

The angle of its deviation from the lateral platform where the largest principal stresses act is $\theta \cong 51°$ and the calculated is $\theta^* = \varphi + \alpha_\xi = 39 + 8° = 47°$.

In the limiting state of stress-strain ratio μ in mutually perpendicular directions tends to the value of 0.8. The results of these tests show that the lateral expansion coefficient of the soil is closely connected to the formation and the angle of deviation of the shear surface. For fine sands, this coefficient is equal to $\mu \cong (0.4 \div 0.5) tg\varphi$ up to the limit state of stress and $\mu \cong (0.8 \div 1) tg\varphi$ in the limiting state of stress (Figure 1.24, b).

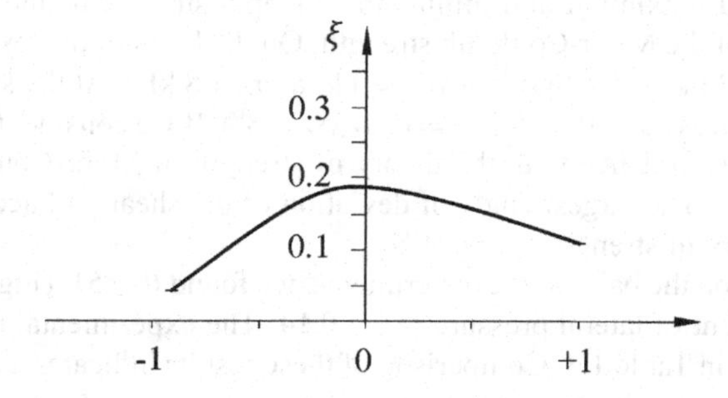

Figure 1.23 Graph of the coefficient of lateral pressure on the type of stress state (−1) – the active pressure; (0) – pure shear; and (1) – passive pressure.

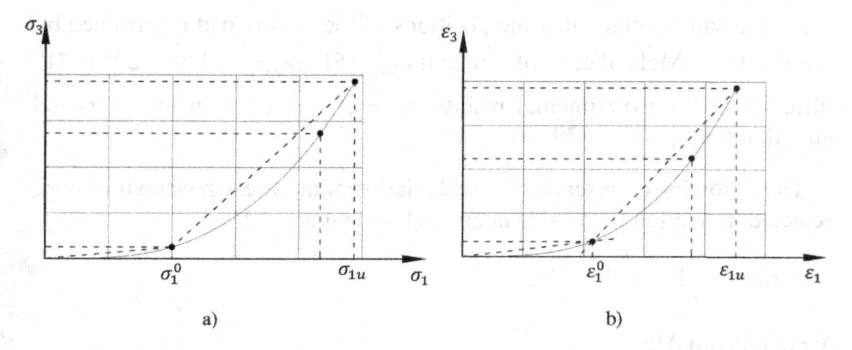

Figure 1.24 Typical plots of changes in the coefficient of lateral pressure in the absence of lateral movements and deformations at possible lateral movements of soil.

Thus, when we use the term "plastic flow of soil," we mean that in this condition the soil as opposed to liquid is characterized by a particular value of the coefficient of lateral pressure $\xi \ll 1$, and lateral expansion coefficient whose value is more than $\mu > (0.4 \div 0.5)tg\varphi$. Therefore, the soil under the limit state of stress known from the literature [17, 19], the ratio corresponding to the deformation theory of elasticity, is not acceptable: $\xi = \frac{\mu}{1-\mu}$.

The ratio of soil movements in a marginal state of stress along the principal axes can be described by the following relationship:

$$\mu = \frac{\Delta\varepsilon_x}{\Delta\varepsilon_z} \cong (0.7 \div 0.8)tg\varphi \tag{1.29}$$

$$\Delta\varepsilon_s = \sqrt{\Delta\varepsilon_x^2 + \Delta\varepsilon_z^2} \tag{1.30}$$

The obtained maximum and minimum principal stresses define the basic parameters of the Mohr-Coulomb strength. On the basis of this experiment, when q = 0.25 kPa, we find that $\sigma_{1\beta} \approx 1$ kPa; $\sigma_3 = 8$ kPa. At the same time, in a state of stress $\mu \cong 0.8$ limit $\xi = 0.14(\alpha_\xi = 8°)$. If we consider the results obtained in accordance with the theory of strength of Mohr-Coulomb, we get $\varphi_M = 51°$. The largest angle of deviation of the shear pad according to Mohr's theory of strength $\alpha_M = 71°$.

In reality, on the basis of the experiments, we found $\theta \cong 51°$ (Figure 1.21). The coefficient of lateral pressure is $\xi = 0.14$. The experimental results are summarized in Table 1.3. Comparison of these results indicates a difference between the strength parameters:

$$\alpha_M - \theta = 71° - 51° = 20°$$

Thus, we can conclude that the positions of the sliding pad determined by the Coulomb-Mohr theory of strength $\varphi_M = 51°, \alpha_M = \frac{\pi}{4} \pm \frac{\varphi}{2} = 20°$ и $71°$ differ from the experimental results, $\alpha_A = \varphi^* = \varphi + \alpha_\xi$; in the proposed equation $\theta^* = \varphi + \alpha_\xi = 39 + 8° = 47°$

The following represent a static calculation of the stress state on the court, rejected in relation to the side at an angle θ. (Figure 1.21)

$$sin\theta = \frac{\tau_\theta}{\sigma_{3\theta}}; \; cos\theta = \frac{\sigma_\theta}{\sigma_{3\theta}};$$

Weight prism ABC

$$G = \frac{a^2 tg\theta * \gamma}{2};$$

And its projection on the vertical surface of the alleged shear AC

$$\sigma_g = \frac{b*\gamma*tg\theta cos\beta\, cos\theta}{2}; \tau_g = \frac{b*\gamma*tg\theta sin\beta\, sin\theta}{2}$$

where $\beta = \frac{\pi}{2} - (\theta + \varphi) = \frac{\pi}{2} - (\alpha_\xi + 2\varphi)$; σ_θ and τ_θ – strength components of the considered platform; σ_g и τ_g – also from gravitational (own weight of an inclined bed thickness *b*) strength. In this case θ is the angle of repose.

Define the normal and tangential strength acting on the considered platform.

$$\sum\sigma_\theta = \sigma_{1\theta} cos^2\theta + \sigma_g \quad (1.32)$$

$$\sum\tau_\theta = \sigma_{1\theta}\frac{sin2\theta}{2} - \tau_g \quad (1.33)$$

$\sigma_{1\theta} = \frac{N_1}{b}$ – maximum end strength

In case of $\beta = 0$, $\theta^* = \varphi + \alpha_\xi$, then

$$\sum\sigma_\theta = \sigma_{1\theta} cos^2\theta + \sigma_g \quad (1.34)$$

$$\sum\tau_\theta = \sigma_{1\theta}\frac{sin2\theta}{2}$$

$$\sigma_g = \frac{b*\gamma*tg\theta\, cos\theta}{2}$$

In accordance with the equation of Coulomb's theory of strength

$$\frac{\tau_\theta}{\sigma_\theta} = tg\varphi \quad (1.35)$$

$$tg\varphi = \frac{\sigma_{1\theta} sin2\theta}{2(\sigma_{1\theta} cos^2\theta + \sigma_g)} \quad (1.36)$$

$$\sigma_{1\theta} = \frac{\sigma_g}{sin2\theta - 2cos^2\theta\, tg\varphi} \quad (1.37)$$

From which it follows that the limiting strength acting on the end of the limiting slope is directly proportional to the magnitude of natural pressure and the angle of internal friction of the soil.

Thus, in the condition of a triaxial strength state, when the direction of the lateral normal strength $\sigma_{3\theta}$ *and the angle of repose of the soil are the same, there is a formation of a new platform shear, deviation from the lateral areas, where the highest principal stresses at an angle is equal to* $\theta^* \cong \varphi + \alpha_\xi$.

Let's define limiting strength at the end of sliding down a slope with an angle of deviation relative to the horizon equal to the angle of repose. The Cartesian coordinate system regards this problem. For this case, tests with the ratio of normal to the inclined surface and the main voltage $\sigma_{1\beta}$ σ_1 are as follows:

$$\sigma_3 = cos\varphi * \sigma_{3\theta u} \tag{1.38}$$

Assuming that the inclined slope of the soil layer is at rest and is not loaded from above, then

$$\sigma_1 = \frac{\gamma b}{2cos\varphi} \tag{1.39}$$

$$\sigma_{1\theta u} = \frac{\gamma b}{2}; \quad \sigma_1 = \frac{\sigma_{1\theta u}}{cos\varphi}$$

$$\xi = \frac{\sigma_{1\theta u}}{\sigma_{3\theta u}} = \frac{\gamma b}{2 * \sigma_{3\theta u}} \tag{1.40}$$

Or ultimate lateral stresses acting in the perpendicular direction of the seam

$$\sigma_{3\theta u} = \frac{\gamma b}{2 * \xi} \tag{1.41}$$

where σ_3 and σ_1 – main strength corresponding to the global coordinate system; $\sigma_{\theta u}$, ξ – respectively the passive resistance of the soil and the coefficient of lateral pressure, determined in vitro when tested in an extremely inclined state; b – the thickness of the reservoir. Equation 1.41 gives additional possibilities to determine the critical load acting in a direction perpendicular to the inclined plane. It can also serve as a basis for determining critical loads on the surface of the soil. These application problems of the theory of strength of soils will be discussed below with reference to the definition of critical loads and the stability of the array of soil, which has the form of a truncated cone.

Example

The goal of the task is to determine the ultimate critical load acting on the end face of the extremely inclined formation $\sigma_{\theta u}$. The thickness of the formation is b = 4 m. The angle of repose is $\varphi = 39°$, the soil density is $\gamma = 14 kH/m^3$ and the lateral pressure coefficient is $\xi = 0.11$. From the equation (1.24) we define

$$\sigma_{3\theta u} = \frac{\gamma b}{2 * \xi} = \frac{14 * 4}{2 * 0.14} = 200\text{kPa}$$

or

$$\frac{\tau}{\sigma} = \frac{\sum H}{\sum V} = \frac{N_3 sin\theta}{N_1 + N_3 cos\theta} = tg\varphi$$

$$N_3 sin\theta = tg\varphi(N_1 + N_3 cos\theta)$$

$$N_3 = N_1 \frac{tg\varphi}{sin\theta - tg\varphi * cos\theta}$$

when $\theta \cong \varphi + \alpha_\xi = 39 + \text{atan}(0.14) = 39^\circ + 8^\circ = 47^\circ$

$$N_1 = \frac{b^2}{2tg\varphi}\gamma = \frac{4^2}{2*0.81}14 = 138 \text{ kN}$$

$$N_3 = 138\frac{0.81}{0.73 - 0.81*0.68} = 630 \text{ kN.}$$

and $\theta = 51^\circ$

$$N_3 = 138\frac{0.81}{0.78 - 0.81*0.63} = 414 \text{ kN}$$

From the presented examples it follows that the position of the angles of deviation of the shear pads significantly affects the ultimate resistance of soils to shear and, accordingly, the lateral pressure coefficient of the soil.

Experiment 3

The purpose of the experiment is to determine the critical load and possible shear surfaces under axial loading of the soil, which has the appearance of a truncated cone.

To determine the limit of the critical load on the ground surface and the shear tests, we conducted the tests with a volume of soil having the form of a truncated cone (Figure 1.26). As noted above, this form of loose soil is stable. This experience is an analogue of uniaxial testing for structured materials. These tests make it possible to define more clearly the strength and stability of loose soils. A distinctive feature of these tests is that we keep the shape unchanged during axial loading of the volume of soil. It is known that during triaxial tests of soil samples shaped into a cylinder, a homogeneous state of stress in the process of deformation is broken.

The selected variant of the limiting deflection angle is always constant, and the onset of a limiting state depends on external forces, and the contact area of the print stamp. This is achieved using a special raised disk mounted on the table. The load to the upper platform of the truncated cone is transmitted using a guide plunger pair attached to the center of the U-shaped frame. At the top of the movable plunger, a rod is provided for the cargo area. On the same rod we attached Measurer

Table 1.3 The results of experimental studies of soil strength. Comparison of parameters of strength and deflection angles of shear pads

Product name, the experimental setup and the type of the stress state	*Limit voltage*			*Angle of repose*	φ_k *form. (2.11 and 1.25) 39*	φ_m *form (2.13 and 1.28)*	φ_T *form (1.28)*	$\varphi_A / \alpha_A /$ *(experiment)*
	kPa	σ_1 *kPa*	$\xi = \frac{\sigma_3}{\sigma_1}$					
1 – flat tray, passive pressure, horizontal tray	9.2	88.57	0.11	36	39	55/72	39/45	39/51
2 – flat tray, a passive pressure tilt tray (Example 3)	1	8	0.13	36	39	51/71	41/45	40/47
3 – flat tray, active pressure, horizontal tray	18.5	230	0.08	36	39	58/29	41/45	41/51
4 – stabilometry of type A (crushing)	100 150 200	375 590 760	0.26 0.24 0.263	36	39	36/63	30/45	39/39
5 – triaxial of type A (crushing, vacuum)	90	320	0.28	36	39	35/63	30/45	39/39
6 – stabilometry of type B (dilation)	460 480	104 104	0.23 0.22	36	39	41/65	32/45	39/39
7 – stabilometry of type B (the condition of compression)	24 102	200 600	0.12 0.17	–	–	–	–	–
Torsion device and tubular samples. Hydrostatic with torsion (2.10) when $\sigma_\theta = 50,100$ and 200 kPa	39	153	0.26	36	39	39/64	σ / τ 50/40	42/42
	70	270	0.26			39/64	σ / τ 100/81	40/40
	140	538	0.26			35/63	σ / τ 200/161	39/39

(measurement) to within 0.01 mm. The load is transmitted to the ground by means of a thin (3 mm) glass plate having a size of 100 × 100 mm. The initial weight of the moving rod and the glass plate is 10 N. The lower radius of the base of the cone is $r_1 = 144$ mm. The upper initial radius's measured fingerprint is $r_2 = 46.3$ mm. The load on the stamp is transmitted by small steps, not more than 2 N. The results of the experiment are shown in Table 1.4.

Suppose that on the surface of a truncated cone there is attached a uniformly distributed load $q = N / A$ (Figure 1.25). We increase the load N until the onset of loss of equilibrium and the transition of the soil to another equilibrium state. This will cause a sharp breakdown, and the stamp will move vertically by an amount S. The new state of equilibrium will occur by increasing the contact area of the print stamp and reducing stress, at the expense of some increase in the bulk density of the soil. Part of the extruded lateral extension of soil will slide to the bottom, beyond the scope of the test. At the same time, the external shape of the ground will remain unchanged; it will be only slightly changed in its geometric dimensions. To determine the critical forces on the surface, you can pre-select the following design scheme:

a) Under the stamp seal forming the core of I in the form of an inverted cone with an apex angle equal to 2α = 2 * (90 – φ). On the surface of the AS wedge in the limiting state, the stresses $-\sigma_\varphi$ and τ_φ act. There is also another pad deviated from this by an angle of (90 – 2φ) (see Figure 1.25). It forms a cone with an apex angle equal to 2α = 2 * φ. The distinguishing feature of this platform is that the outside is formed from another volume of soil, which in cross section has the shape of an inclined plane limit. This amount of soil on the side surface of the load takes the normal passive. In the limiting state of stress there is formed shear surface BD (Figure 1.25), which takes a passive earth pressure from the surface

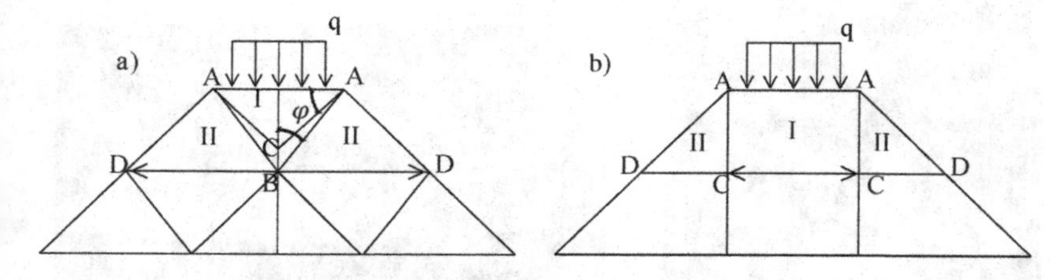

Figure 1.25 Schemes of the state of stress and the shear surface in the elementary volume of soil: a) the design scheme with a central core of compacted I and persistent side prisms of triangular cross-section II; b) the same as with the central cylindrical part and two rectangular lateral abutment prisms II.

AB. The stresses σ_θ and τ_θ act on the wedge surface AC and AB (Figure 1.21). Given that the value of the site AB values are $\tau_\theta = \min$ and $\sigma_\theta = \max$, it can be called the principal. This wedge is supported by two inverted side prisms II.

b) A frustoconical shape (truncated cone) is conditionally divided into two elements: a cylinder I loaded with the top die, and two lateral triangular abutments II.

Based on the results of the above experiments, it can be assumed that a loaded top stamp pad can be distinguished by characteristic deviations from the horizontal at an angle (90 – φ). Although the condition of the ultimate state is not formally followed on this pad, but the maximum normal stresses act on it and, for this reason, the soil shear surfaces are deviated with respect to the slope by the angle $\theta \cong \varphi$. This is confirmed by an analysis of the trajectory of the movement of special beacons which are made of shortened matches and installed along the lateral surface of the slope (Fig. 1.26). Thus, in the section under the stamp, it is conditionally possible to distinguish a

Figure 1.26 a) General view of the test bed; b) the testing process; c) diagram of the formation of a densified core and shear surfaces.

central wedge with an angle at the apex of 2α = 2 * φ and two quadrangular and triangular lateral thrust prisms in the section (Fig. 1.26 c). Family shear pads parallel to the surface of the slope are between AD and an angle (180 – 2φ) and (2φ). The ultimate state of the inclined thrust of quadrangular prism from the action of normal passive stress was considered above, by the example of experiment No. 2.

To determine the critical load acting on the surface of soil in the form of a truncated cone, we will design a scheme of interaction of inverted pyramid I triangular and rectangular lateral thrust prism II (Figure 1.26 c).

Calculation by the first scheme

Basic geometric shapes: the height of the wedge:

$$h_{AC} = \frac{r_1}{tg\varphi} = \frac{D}{2tg\varphi} = r_1 / f \quad (1.42)$$

The area of the loaded surface:

$$S_{r_1} = \pi * r_1^2 \quad (1.43)$$

The volume of a truncated cone:

$$V_{I,II} = \frac{1}{3}\pi h\left(r_1^2 + r_2 r_1 + r_2^2\right) = \frac{1}{3}\pi h r_1^2\left(1 + \frac{1}{\sin_\varphi^2} + \frac{1}{\sin_\varphi^4}\right) \quad (1.44)$$

The volume of hard triangular prism II:

$$V_{II} = \frac{1}{3f}\pi r_1^3\left(1 + \frac{1}{\sin_\varphi^2} + \frac{1}{\sin_\varphi^4}\right) - \frac{1}{3f}\pi r_1^3 = \frac{1}{3f}\pi r_1^3\left(\frac{1}{\sin_\varphi^2} + \frac{1}{\sin_\varphi^4}\right) \quad (1.45)$$

The volume of hard quadrangular prism II:

$$V_{II} = \frac{2}{3f}\pi r_1^3\left(\frac{1}{\sin_\varphi^2} + \frac{1}{\sin_\varphi^4}\right) = \frac{2\pi r_1^3}{3f\sin_\varphi^4}\left(\sin_\varphi^2 + 1\right) \quad (1.46)$$

Define normal force directed onto the surface of AB. For this composition the equilibrium equation of the forces in relation to surface AB is:

$$N_{AB} = (N_1 + g) * \sin\varphi \quad (1.47)$$

or:

$$\frac{N}{\pi * r_1^2}\sin\varphi = \left(\sigma_{1u} + \frac{g}{\pi * r_1^2}\right) * \sin_\varphi^2 \quad (1.48)$$

where $\frac{\pi * r_1^2}{\sin\varphi} = \Omega_s$ – the surface area of the side AB persistent prism; $\sigma_{1u} = \frac{N_1}{\pi * r_1^2}$ – limiting the voltage on the surface I; g – weight soil core (ABA);

$$\frac{N_{AB}}{\Omega_s} = \sigma_{1AB} = \left(\sigma_{1u} + \frac{g}{\pi * r_1^2}\right) * \sin_\varphi^2 \tag{1.49}$$

We define the ratio limit of the forces acting on the inclined prism of stubborn II (Figure 1.25). To determine the limit of critical load N_{AB} acting on the thrust side of inclined prism II, we will use the results of experiment 2.

$$N_{AB} = \frac{N_g^{II} \cos\varphi}{\xi}; \; N_g^{II} = \frac{2\pi\gamma r_1^3}{3 * \sin_\varphi^4 tg\varphi}\left(sin_\varphi^2 + 1\right);$$

$$N_{AB} = \frac{\pi\gamma r_1^3 \cos\varphi}{k_1 * \xi * \sin_\varphi^4 tg\varphi} \tag{1.50}$$

$$k_1 = \frac{3}{2\left(sin_\varphi^2 + 1\right)}$$

or

$$\sigma_{1AB} = \frac{N_{AB}}{\Omega_s} = \frac{\cos\varphi \sin\varphi \pi\gamma r_1^3}{k_1 \pi \xi * r_1^2 \; * \sin_\varphi^4 tg\varphi} = \frac{\gamma r_1}{k_1 \xi} * \frac{\cos\varphi}{\sin_\varphi^3 tg\varphi} \tag{1.51}$$

where ξ – the coefficient of lateral earth pressure in the limiting condition, as measured by the test results in the oblique state soil stratum.

Taking into account (1.49), we finally define the limit stress acting on the surface of a truncated cone:

$$\xi * \left(\frac{N_1}{S_{r_1}} + \gamma * h_{AC}\right) = \frac{N_2}{S_{r_2}} f \tag{1.54}$$

The left side of equation (1.47) has a lateral thrust of a centrally loaded cylinder of height h_{AB}, and the right side has passive horizontal resistance of a persistent lateral rectangular prism II.

Solving the equation (1.54) with respect to N_1, and taking into account (1.42), we obtain:

$$N_1 = N_2 \frac{S_{r_1}}{S_{r_2}} \frac{f}{\xi} - \gamma\pi f r_1^3 = \frac{4}{9} \frac{\pi f^2 \gamma r_1^3}{\xi} - \gamma\pi f r_1^3 \tag{1.55}$$

$$\text{Let } M_{f2} = \pi * f \text{ и } M_{f1} = \frac{4}{9}\pi f^2 \tag{1.56}$$

Taking into account (1.56), we have:

$$N_{1u} = M_{f1} r_1^3 \frac{\gamma}{\xi} - M_{f2} \gamma r_1^3 = \gamma r_1^3 \left(k \frac{M_{f1}}{\xi} - M_{f2} \right) \tag{1.57}$$

where –k ≅ 5 is a correction factor.

Equation (1.57) determines the value of a critical load installed on the soil surface in the form of a truncated cone with a print area of a circle with radius r_1.

If the equations (1.53) and (1.55) as the value of the coefficient of lateral pressure ξ substitute value corresponding to the condition of rest, the N_{1R} will match the initial critical load, and vice versa, with the critical value, we obtain the maximum critical load N_{1u}.

Equation (1.57) can also be written as critical normal stresses:

$$\sigma_{1u} = \frac{N_1}{\pi r_1^2} = k \frac{4}{9} \frac{f^2 \gamma}{\xi} r_1 - \gamma f r_1 = \left(k \frac{M_{f3}}{\xi} - f \right) \gamma r_1, \tag{1.58}$$

where $M_{f3} = \frac{4}{9} f^2$.

According to the results of such tests, we can also determine the rate and magnitude of lateral pressure corresponding to the start of plastic deformation. This value is determined by solving the equation (1.23) with respect to ξ:

$$\xi = k \frac{M_{f2} r_1^3 \gamma}{N_{1u} + M_{f1} \gamma r_1^3} \tag{1.59}$$

Example 1

Calculation of critical load acting on the upper platform of the soil mass in the form of a truncated cone (for example, experiment 3). Background: the angle of repose – 36°; internal friction angle φ = 39°; soil density $\gamma = 1.48\ г/cm^3$. The lower and upper radius of the base of the cone $R_{1,2}$ = 14.4; 4.64 cm; ξ = 0.16 and k = 5. Results are presented in Table 1.4. Experimental results and analytical solutions coincide satisfactorily on the first circuit and, to a lesser degree, a second calculation circuit.

Table 1.4 The results of calculation of the critical load, acting on the upper platform of the soil mass in the form of a truncated cone

Load	*Sludge*	*Change in the radius of the print* r_1	*The voltage on the surface of a truncated cone*		
			In the experiment	*Calculation of the first scheme*	*Calculation of the second scheme*
N	*mm*	*cm*	*kPa*		
1	0	4.63	1.5	2.1	1.0
1.1	0.5	4.64	1.6	2.5	1.0
1.2	3.6	4.69	1.7	2.5	1.0
1.3	7.5	4.75	1.8	2.5	1.1
1.4	12.9	4.83	1.9	2.6	1.1
1.5	13.9	4.84	2.0	2.6	1.1
1.6	20.1	4.94	2.1	2.6	1.1
1.7	21.3	4.95	2.2	2.7	1.1
1.8	26.8	5.04	2.3	2.7	1.1
2	34.8	5.16	2.4	2.7	1.2
2.2	36.9	5.19	2.6	2.8	1.2
2.2	43.	5.28	2.5	2.8	1.2
2.4	53.6	5.44	2.6	2.9	1.2
2.8	59.8	5.54	2.9	2.9	1.2
3	72.4	5.73	2.9	3.0	1.3
3.5	83.7	5.90	3.2	3.1	1.3
4	94.	6.06	3.4	3.2	1.4

The task: Plane (flat) deformation

Calculation by the first scheme: let the width of the loaded surface be equal to *2b*. $N_1 = 2bg$ Make equilibrium equation of forces to the surface AB:

$$N_{AB} = \frac{1}{2}(N_1 + 2G) * \sin\varphi \tag{1.60}$$

or $$\frac{N_{AB}}{b}\sin\varphi = \left(\sigma_{1u} + \frac{G}{b}\right) * \sin^2_{\varphi}$$

where $\frac{b}{\sin\varphi} = \Omega_s$ – the surface area of the side AB persistent prism; b – the half-width of the loaded surface; σ_{1u} – the limiting strength at the surface I; G – half the weight of the compacted core area (AB);

$$\frac{N_{AB}}{\Omega_s} = \sigma_{1AB} = \left(\sigma_{1u} + \frac{G}{b}\right) * \sin^2_{\varphi} \tag{1.61}$$

We define the balance of forces acting on the inclined thrust prism II (Figure 1.25);

$$N_{AB} = \frac{N_g^{II}}{\xi} cos\varphi N_g^{II} = V_{II}\gamma = \frac{b^2\gamma}{\sin_\varphi^2 tg\varphi} N_{AB} = \frac{b^2\gamma cos\varphi}{\xi k\sin_\varphi^2 tg\varphi} \quad (1.62)$$

From equation (1.26) it follows that

$$N_1 = \frac{2N_{AB}}{\sin\varphi} - 2G = \frac{2b^2\gamma cos\varphi}{\xi k\sin_\varphi^2 tg\varphi} - \frac{b^2\gamma}{tg\varphi} = \frac{b^2}{tg\varphi}\gamma\left(\frac{M_{f1}}{k*\xi} - 1\right) =$$

$$= bh\gamma\left(\frac{M_{f1}}{k*\xi} - \frac{1}{2}\right) \quad (1.45)$$

where N_g^{II} – the weight of the thrust side of the prism; k – coefficient;

$$\mathrm{M}_{f1} = \frac{2\cos\varphi}{\sin_\varphi^2}; \mathrm{h} = \frac{\mathrm{b}}{\mathrm{tg}\varphi} \quad (1.63)$$

Taking into account (1.61), we finally define the limit stress acting on the surface of the prism inclined thrust II:

$$\sigma_{1AB} = \frac{N_{AB}}{\Omega_s} = \frac{b^2\gamma cos\varphi}{\xi k\sin_\varphi^2 tg\varphi}\frac{\sin\varphi}{b} = \frac{b\gamma cos\varphi}{\xi k\sin_\varphi tg\varphi};$$

$$\sigma_{1u} = \frac{\sigma_{1AB}}{\sin_\varphi^2} - \frac{G}{b} = \frac{\sigma_{1AB}}{\sin_\varphi^2} - \gamma h_{AB} = \frac{b\ \gamma cos\varphi}{\xi k\sin_\varphi^3 tg\varphi} - \frac{\gamma b}{2tg\varphi} =$$

$$= \frac{\gamma b}{2tg\varphi}\left(\frac{2cos\varphi}{\xi k\sin_\varphi^3} - 1\right) = \frac{\gamma b}{2tg\varphi}\left(\frac{M_{f1}}{k*\xi} - 1\right) \quad (1.64)$$

Calculation of the second scheme

We form the equation of balance of powers in relation to the X axis:

$$\xi * \left(\frac{N_1}{S_{r_1}} + \frac{\gamma * h_{AC}}{2}\right) = \frac{2N_2}{S_{r_2}} f \quad (1.65)$$

$$N_1 = b * g$$

The left side of equation (1.65) has the meaning of lateral thrust of the centrally loaded rectangle of height h_{AB} (1.42), and the right side has passive horizontal resistance of the persistent lateral rectangular prism II.

Solving the equation (1.65) with respect to N_1 and taking into account (1.11), we obtain:

$$N_1 = 2N_2 \frac{S_{r_1}}{S_{r_2}} \frac{f}{\xi} - \gamma \frac{b^2}{2tg\varphi} = \frac{2f}{2\xi} \frac{b^2 \gamma}{tg_{\varphi}^3} \frac{b}{b} tg_{\varphi}^2 - \gamma \frac{b^2}{2tg\varphi} =$$

$$= \frac{2b^2\gamma}{2\xi} - \gamma \frac{b^2}{2tg\varphi} = \frac{b^2\gamma}{2}\left(\frac{2}{k\xi} - \frac{1}{tg\varphi}\right)$$

$$h = \frac{b}{tg\varphi}; \; S_{r_2} = S_{r_1} = b \tag{1.66}$$

Equation (1.66) is used to determine the value of the critical load installed on the surface of soil slope of width b, while the surface load intensity is q.

$$\sigma_{1u} = \frac{N_1}{b} = \frac{b\gamma}{2\xi} - \gamma \frac{b}{2f} = \frac{b\gamma}{2}\left(\frac{2}{k\xi} - \frac{1}{f}\right) \tag{1.67}$$

Example 2

Calculation of critical load acting on the upper platform of the ground slope. Background: lateral pressure coefficient $\xi = 0.4$, angle of repose – 400; internal friction angle $\varphi = 40°$;

$$b = tg\varphi \; h = 0.84 * 1 = 0.84 \text{m}$$

The critical load to the ground by the formula (1.45):

$$N_1 = bh\gamma\left(\frac{M_{f1}}{k * \xi} - \frac{1}{2}\right) = 0.84 * 1 * 19\left(\frac{3.7}{2 * 0.4} - \frac{1}{2}\right) = 66kN$$

$$g = \frac{N_1}{b} = \frac{66}{0.84} = 78 \text{ kPa}$$

Next, the critical load to the ground by the formula (1.66)

$$N_1 = \frac{b^2\gamma}{2}\left(\frac{1}{k\xi} - \frac{1}{tg\varphi}\right) = \frac{0.84^2 19}{2}\left(\frac{2}{2 * 0.4} - \frac{1}{0.84}\right)$$

$$= 6.7 * 0.06 = 9kN$$

$$g = \frac{N_1}{b} = \frac{9}{0.84} = 11kPa$$

The results of these examples show that the first settlement scheme is closer to reality (the results of the experiment). Therefore, it can be used as a basis for the calculations for the first limit of the soil.

Conclusions for Chapter 1

1 The position of the sliding surface is of great practical importance, and it greatly influences the accuracy of determining the limit values of forces: active and passive pressure.
2 Traditionally, in the calculations, the orientation of the shear pads according to the Coulomb-Moore theory of strength is determined by the equations (1.2). However, our studies have shown that the deflection angles of surfaces offset in relation to areas where there are maximum and minimum principal stresses are equal to $\theta^* \cong \varphi + \alpha_\xi$ и $\theta^* \cong \frac{\pi}{2} - (\varphi + \alpha_\xi)$. At the same time, the divergence angles of these areas, with active and passive loads, are (90° – 2φ).
3 It is traditionally believed that during flat shear tests in soils (sand), immediately after its horizontal loading, the process of loosening of the soil occurs. According to our research, the rise of the upper die shear is associated with the direction of movement of soil, slanting upward (Figures 1.2 and 2.1). The experiments on torsion devices with fine sands confirm these conclusions (Figure 1.16). From the beginning of deviatoric loading, the soil is almost always sealed (to a greater extent for loose and to a lesser extent for dense soil). As the soil approaches its ultimate state, the consolidation stabilizes and by the beginning of the growth of the plastic flow its loosening is observed.
4 Conducting experiments, we found that the strength properties of soils and surface shear deflection angles obtained by the theories of Mohr-Coulomb and Tresc Hill differ from the actual results of the experiments. Otherwise, for soils, the ratios of the main ultimate stresses, the direction of the shear pads and the magnitude of the normal and tangential stresses acting on them obey other laws. The definition of these fundamental values of Mohr circles is not really observed in practice. Therefore, thus obtained calculation results are inaccurate (see Table 1.3).
5 The experiments performed under conditions of triaxial compression of the soil revealed that the transverse strain coefficient and the transverse expansion coefficient depend on the deviator stresses. In particular, it was found that lateral displacements during initial axial loads do not develop as intensively as axial ones. This situation is approaching the condition of compression (Figure 1.24). With a further increase in axial stresses, lateral deformations begin to appear, and as the soil approaches the point of destruction, it begins

to grow rapidly. From this point on, the intensity of the lateral deformation begins to exceed the vertical. For this reason, at the approach to the critical stress state, primer begins to increase in volume and disintegrates.

6 Experiments on the pattern of passive and active earth pressures found that in contrast to the angle of internal friction, the coefficient of lateral soil pressure is not constant and depends on many factors: the amount of displacement and the level of the stress state; the orientation angles of the shear areas; and so on. For this reason, the value should be determined in accordance with the terms of its physical nature. For example, when evaluating the stress state of the soil slope stability, it is necessary to determine the value in testing without the possibility of lateral expansion. In solving problems of strength and stability, including nonlinear problems, this characteristic should be determined subject to the possibility of lateral expansion of soil.

7 According to the results of determining the critical load during axial loading of the soil bulk, having the form of a truncated cone, it was found that such tests differ from triaxial stabilometric tests. As is known, in the last test, in the process of loading, a strong distortion of the shape occurs, which does not allow adjust clear criteria of destruction. The selected variant of the limiting deflection angle is always constant, and the onset of the limiting state depends only on the external force and the contact area of the print stamp. Such tests allow a comparison of ultimate loads on the soil and the ultimate strength and stability of soils.

8 On the basis of comprehensive studies of soils was found the analytical solution of determining the value of the coefficient of lateral pressure for the critical load (stress) on the soil in the form of a prism and the form of a truncated cone. In the case of testing the soil in the form of a truncated cone in the uniaxial compression, the preferred design scheme is shown in Figure.1.25a. In this case, the calculation results coincide satisfactorily.

2 Strength of the soil

2.1. Discourse on the surface and shear strength parameters of soil

To highlight this issue, let us consider the condition of the surface strength of repose. If on the surface slope we highlight the elementary volume weight of soil G and external force P, the surface potential shear F platform has two oppositely directed forces:

$$T = (G + P)sin\beta \text{ и } T_u = (G + P)cos\beta \cdot tg\beta + cF \tag{2.1}$$

here β – angle of repose; c – the gearing power; $F = F^H cos\beta$ – sloping sliding surface area; F^n – its horizontal projection.

In the context of the limiting state of stress and lack of lateral pressure $\sigma_3 = 0$ angle of repose $\beta = \varphi^* - atan\xi$. These two forces are balancing and satisfying the following condition:

$$(G + P)sin\beta = (G + P) \cdot cos\beta tg\varphi + cF \tag{2.2}$$

or

$$tg\beta = tg\varphi + \frac{cF}{(G + P)cos\beta} \tag{2.3}$$

Equation (2.3) and the parameters for cohesionless soils have the following physical nature. It follows from the addition of natural soil as sedimentary rocks. The parameters in this equation, although they have a sense of friction and cohesion, are in fact still mathematical parameters of the generalized equations of the theory of Coulomb strength. For example, the strength parameter "φ" of the soil differs from friction on the ideal surface of the mineral against the mineral (figure 2.3), and "c" from the structural bond intrinsic for hardened solids. Here, the φ is mainly characterized, as many have noted

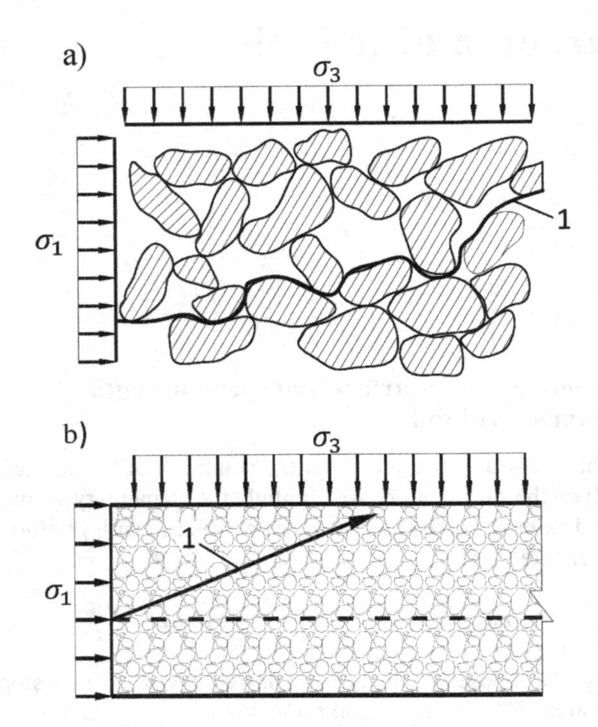

Figure 2.1 The scheme of forming the inclined surface shears in the sands of the passive tense condition 1 – surface of shear.

in the literature [5, 8, 9], by the conditions of contact between the mineral particles, density, size and so on. The second parameter depends more on the shape and size of the mineral particles. For example, for a plate, needle and prismatic minerals it has a different measure of resistance. The chemical composition of natural cements that bind mineral particles has a significant effect on clay soils. Therefore, this characteristic – to a lesser extent of sand and more for salty and clay soils – greatly depends on its humidity.

Although the equation (2.3) is described mathematically in the same way as for structured soils (for example, for clay soils), physically the second parameter has a different meaning. For clay soils with a particle size less than 0.005 mm, the coupling parameter c characterizes the strength of the structural relationships (natural salts and cement) on the shear.

Thus, the parameter φ for unstructured soils is some integral quantity which characterizes the friction between the particles. This feature depends on the complex properties of soils, such as the mineral composition of the

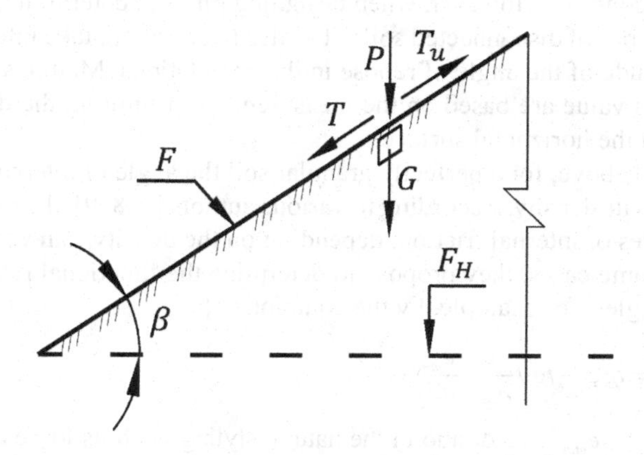

Figure 2.2 Scheme of restraints and shear forces of the elementary volume of soil at the surface of the slope.

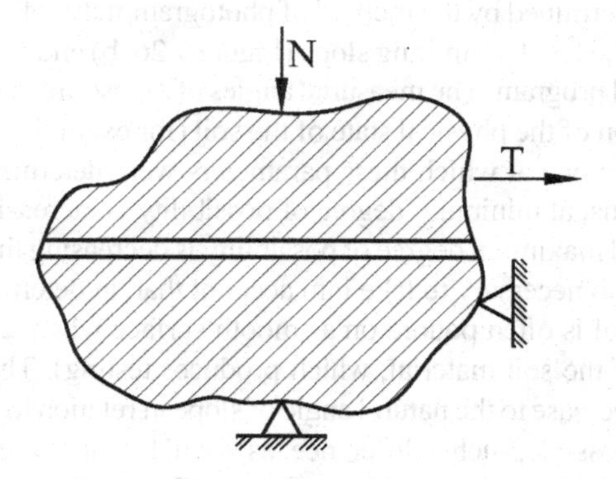

Figure 2.3 Scheme of determination for the true coefficient of friction of the mineral.

rock (hardness), degree of weathered clastic rocks, sizes and surface of roughness (degree of roundness) and so on. Often, for cohesionless soils, the parameter φ is taken as the angle of repose. We should note that the angle of repose is usually smaller than the angle of internal friction of soil. This is because the contact surface of the particle lying on the slope surface is always smaller than the contact area with the same particles lying within the array. On the other hand, the contact surface between the particles is due to the bulk normal stresses. Therefore, the angle of internal friction of soil is not constant and depends on the volume of normal stresses. In practice, this characteristic in the range of normal stress does not exceed 300–400 kPa, taken as a constant. However, when designing pits and determining the stability of slopes of disconnected soils, it is also necessary to take into account the magnitude of the angle of repose in the calculations. Methods of determining this value are based on the measurement of limiting the deflection angle from the horizontal surface.

As noted above, for a perfectly granular soil the angle of internal friction depends on its density. According to various authors [5, 8, 9], the real values of the angles of internal friction, depending on the density, can vary from 2 to 30. In some cases, they propose to determine the functional relationship of these angles, for example, by the equation [3]:

$$tg\beta o = tg\varphi - tg\theta \frac{e - e_{min}}{e_{max} - e_{min}} \tag{2.4}$$

where e, e_{max}, e_{min} – void ratio in the natural styling, with its loose and dense state; $tg\theta$ – correction factor. For example, our studies with fine sands in dry air showed that the difference in the angle of internal friction of the shear tests and that determined by the angle of repose is not more than $3 \div 4°$. The angle of natural slope was determined by the method of photogrammetry of the profile of the poured cone to the level of limiting slope (Figure 1.26, b) and its approximation using a graphical program. The measured angles of repose are $\beta = (0.93–0.94)\ \varphi$. But any deviation of the physical state of the soil (for example, density, humidity, etc.) from the one at which these parameters were determined can lead to various deviations, at minimum degree of possibility is increasing the angle of natural slope and maximum degree of possibility is decreasing the angle of natural slope. It is also necessary to take into account that, in determining the angle of repose, the soil is often poured on a smooth surface where $\varphi_s < \varphi$ (φ_s – the friction angle of the soil material, which produces testing). This situation can also affect the decrease to the natural angle of slope in relation to the actual angle of slope. Most possible, such a difference, as it will be shown below, is affected by the value of the lateral pressure coefficient. The coefficient of lateral pressure on the surface of the slope, where large displacement $\xi = max$, and within the array, where such movement is limited, has a minimum value (Figure 2.2). Perhaps for this reason, there occurs a difference between the angle of internal friction of the soil and the coefficient of repose.

Now we are going to analyze the physical meaning of the earth resistance at low normal stresses. For this, both sides of the equation (2.2) are divided by the contact area:

$$\frac{(G+P)}{F} sin\beta \le \frac{(G+P)}{F} cos\beta tg\varphi + c,\ \tau_\beta \le \sigma_\beta tg\varphi + c \tag{2.5}$$

or

$$\tau_{\beta P} \le \sigma_{\beta P} tg\varphi + \left[\left(\sigma_{\beta g} tg\varphi - \tau_{\beta g}\right) + c\right] =$$

$$=\left(\sigma_{\beta P}+\sigma_{\beta g}\right)tg\varphi+\left(c-\tau_{\beta g}\right) \tag{2.6}$$

where

$$\tau_{\beta}=\frac{(G+P)}{F}sin\beta;\ \sigma_{\beta}=\frac{(G+P)}{F}cos\beta.$$

This famous equation describes the limiting resistance to soil shear strength corresponding to the theory of Coulomb. Equation (2.6) can be represented as follows:

$$\tau_{\beta P}\leq\left[\sigma_{\beta P}+\left(\sigma_{\beta g}+\frac{c-\tau_{\beta g}}{tg\varphi}\right)\right]tg\varphi \tag{2.7}$$

or

$$\sigma_{g}^{1}=\sigma_{\beta g}+\frac{C}{tg\varphi}-\frac{\tau_{\beta g}}{tg\varphi} \tag{2.8}$$

where $\sigma_{g0}=\frac{C}{tg\varphi}$ – the pressure of the real connectivity, or, for a unit area, which characterizes the force equivalent to separation, the vector of which is directed perpendicular to the shear surface; σ_{g}^{1}– is the stress equivalent to the adhesion forces for loose soils and adhesion for clay soils.

Substituting (2.8) for (2.7) we get:

$$\tau_{\beta P}\leq\left(\sigma_{\beta P}+\sigma_{g}^{1}\right)tg\varphi \tag{2.9}$$

Equations (2.7 and 2.8) offer a very convenient way to solve practical problems of the theory of strength and stability of structured soils, which will simplify the math.

Often, when solving engineering problems, soil is conventionally accepted as an infinitesimal element related to a solid deformable body, for example, a cube or a prism affected by stresses (Figure 2.4). In accordance with the theory of mechanics of a solid deformable body, the set of normal and tangential stresses acting across all differently oriented areas will correspond to a stressed state in an elementary volume (point). Graphically, such a stressed state at a point is displayed by Mora circles, which are widely used in soil mechanics. Such an accepted position conditionally or formally makes it possible to analyze the stress state in an elementary volume of the soil as a continuous inseparable medium. This provision was the theoretical basis for determining the most dangerous shear site. As was shown above, for soils, the actual orientation of the shear sites differs from that calculated according to the Coulomb-Mohr theory (π / 4 ± φ / 2) and corresponds, as our studies showed (see Chapter 1.3), to φ* and (π / 2 – φ*), where φ* > φ.

Such a discrepancy between the real and calculated deflection angle of the most dangerous shear site gives grounds to continue research on the strength model of the triaxial-loaded soil. Such a model of strength proposed by the authors of this book is discussed below.

2.2. The theory of soil strength

As we have already noted above, the state of stress at a point, the definition of the parameters of soil strength and the concept of the largest deflection angle is largely dependent on the received coordinate systems. Therefore, any deviation from this system may lead to misinterpretation of the true meaning of the concept of strength. To clarify this, it is necessary to find out the correct term strength of the soil, depending on the stress state of the soil. To clarify this point in the array of soil, isolate a conditional amount of soil in the form of a prism. We regard the state of stress in the form of a prism element of the complex influences of external and internal gravitational forces. It is known that 9 stress components (three normal and six tangential) act on this element (Figure 2.4.). If we consider the task

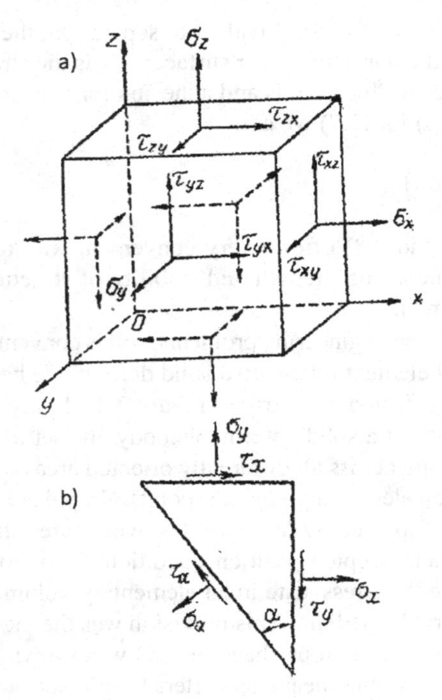

Figure 2.4a The stress state in the elementary volume of the soil is a) spatial, b) the plane problem.

Figure 2.4b Situ plate tests of soils. 1-plate; 2-jack; 3-dynamometer; 4-deflection meter Maximov; 5-frame; 6-rail

of plane deformation, then the elementary volume of unit thickness is taken as the basis, the fixed two rigid plates bounded on both sides. In this case, two normal and two tangential stresses act on the soil. Intermediate normal stress σ_2 in this case it is unchanged and, as a rule, is not involved in calculations (Figure 2.4 b).

Therefore, for the experimental determination of the strength properties of soil, you must adhere to the following rules:

1 *The definition of soil strength, having simple geometric shapes:*

a The strength of the soil in the form of a rectangular prism (cube) (Figure 1.2, 2.4a), which operates a system of stress $\sigma_1 > \sigma_2 > \sigma_3$;

b The strength of the soil in the form of a rectangular prism (cube) (Figure 1.2, 2.4. A), which is affected by some stresses $\sigma_1 > \sigma_3$;

c The strength of the soil in the form of a thick plate (circular, square) (Figure 1.4) with a predetermined area. It is assumed that the calculated shear surface, which is affected by some stresses $\tau \geq 0$ and $\sigma \geq 0$, is parallel to the surfaces of the fronts of the soil sample.

2 *The definition of other forms of soil strength, differing from the rectangular discussed in paragraph a:*

a The strength of the soil, which has a cylindrical shape (Figure 1.6, and 2.6), which operates the system voltages $\sigma_1 > \sigma_{2=3}$;

b The strength of the soil in the form of a thick-walled pipe (Figure 1.17), which operates the system voltages $\tau \geq 0$, σ_θ, σ_r and σ_z.

3 *Determination of strength in soil models imitating its real geometric shapes:*

a Determination of soil strength under plate tests with external pressures N (Figure 2.5);

b Determination of soil under rotational shear N, T and M;

c Determination of the soil in the hole (press metric test) when exposed to normal external radial stresses σ_r.

In our view, when we talk about defining the strength of the soil, it is necessary to adhere to the methods specified in the first and second groups of tests, where a given state of stress is observed. It is necessary to strictly adhere to the postulate of destruction, where constancy of the principal stresses and the orientation of the shear areas is observed. Under the term, the strength of the soil is associated as a shear inside the solid along an inclined platform (parallel inclined surfaces along which plastic movements of the soil that do not fade over time). On this pad, the limiting ratio of normal and shear stresses acts.

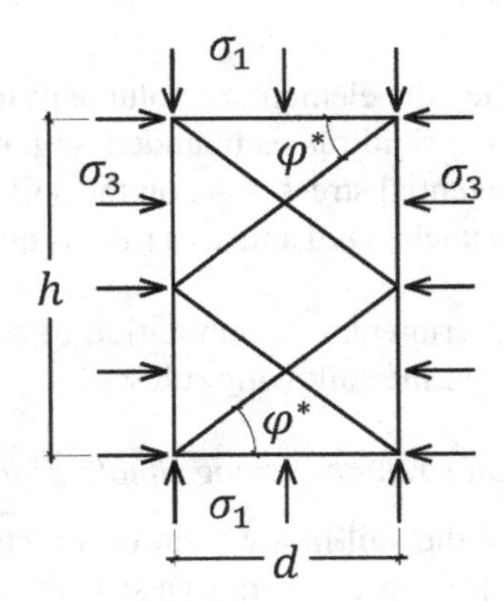

Figure 2.5 Scheme to select the aspect ratio of the soil sample.

Moreover, during the tests, it is necessary to adhere to the situation when, under a critical stress state, a condition was created for the continuous plastic flow of soil layers along the shear pads. Otherwise, when a critical stress state is reached, the soil goes into another equilibrium state, which differs from the original pad. Thus, the conditional volumetric destruction of the soil at a point can be characterized as a result of plastic shears along inclined sliding platforms under conditions of the ultimate stress condition (Figure 2.6).

For an analytical description of the relationship of the concept of strength of soil by the action of volume forces $\sigma_1 > \sigma_2 > \sigma_3$ and on platforms and shear $\tau_\theta \geq 0$ and $\sigma_\theta \geq 0$, use different theory (Coulomb, Coulomb-Mohr, Hill, Mises, Botkin, etc.). The meaning of all these theories is that some specific stress ratio $\tau_\theta / \sigma_\theta$ and orientation angle θ shear platforms are used to estimate the fracture criterion of soil. The problem is that some of these models more accurately assess and predict the strength and stability of the ground in the solution of specific problems of geotechnical engineering.

It is believed that the shear pads direction depends only on the angle of internal friction and is independent of the body forces. But in fact it is not. The fact is that at high values of volumetric forces, its density also increases, and hence the value of the angle of friction. But this can seriously complicate the solution of the task, therefore, given the insignificant change in the value of the angle of friction, when solve practical engineering tasks, it is taken as an averaged constant figure.

Now, we give special attention to the shape and size ratio of the elementary volume of soil. The soil sample must comply with such a size so that you can get the exact characteristics of the strength parameters of the soil. It is necessary to take into consideration that the accepted dimensions of an elementary soil sample do not interfere with the free movement of shear surfaces. Otherwise, the resulting strength characteristics of soils may differ materially from the actual characteristics observed in nature. With this in mind, the main dimensions of the soil sample during the test must meet the following conditions. The ratio of the length and the smaller side (e.g. diameter) of the soil sample should satisfy the following condition (Figure 2.5):

$$h > \frac{d}{tg\varphi} \tag{2.10}$$

wherein *h*, *d* – respectively the height and width (diameter) of the sample. For samples of unit width, for example, fine $\varphi = 39° tg\varphi = 0.81$

$$h > \frac{1}{0.81} = 1.24$$

for $\varphi = 25°$, $tg\varphi = 0.47$

$$h \cong \frac{1}{0.47} = 2.12$$

Thus, to unify the aspect ratio of the soil sample for sandy, sandy loam and light loamy soils, it is usually accepted a value of no more than $\frac{h}{d} = 2$. Figure 2.7 and 2.8 shows the stress condition and position of the surfaces of the shear of a soil sample after compression and tension on triaxial

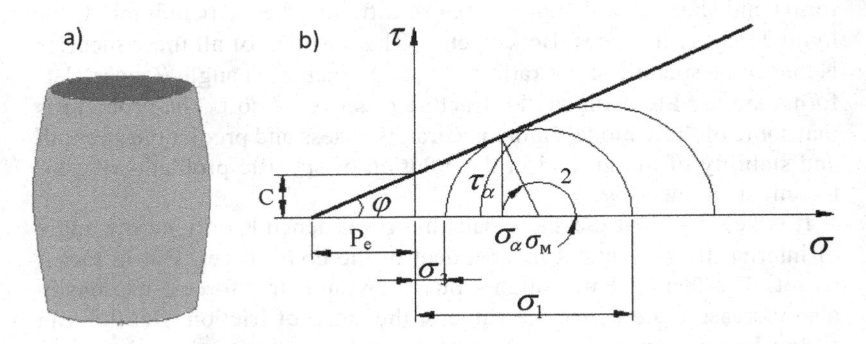

Figure 2.6 Type of plastic consistency of the soil sample after the punching test (a) and Mohr diagram for determining the parameters of soil strength (b).

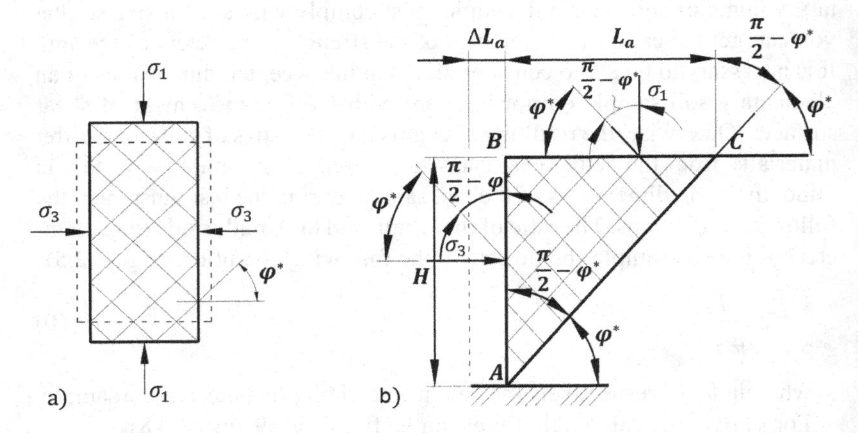

Figure 2.7 Stress state and the position of the shear surface with the active resistance of the soil compression: a) triaxial stress state of the soil sample, and b) an active soil pressure on the fence.

compression devices and comparing it with the stress state of a vertical slope with active and passive soil resistances. The condition of soil strength at shear (case *1c*) corresponds to the Coulomb strength theory and is constructed in a rectangular coordinate system $f(\sigma, \tau)$:

$$f^* \equiv \tau - \sigma * f = 0 \tag{2.11}$$

This condition is graphically outlined in the rising straight line in the coordinate system τ, σ, with slope equal to $f = tg\varphi$. In accordance with this theory, the condition of equality between the angle of orientation of the shear pads with the angle of internal friction of the soil is met. In fact, in tests of soil on devices of one-flat shear, the real surface changes slightly, but deviates from the horizontal. For this reason, as mentioned above, with an increase in horizontal movement of the top ring of the device there has been some rise in the vertical stamp (Figure 2.1). Often it is perceived as a loosening of the soil in a plane offset that in effect is an effect of oblique sliding soil. The nature of the vertical rise of the stamp with the shear tests will be discussed below. In practice, when determining soil strength parameters, stress components are taken at the shear pad parallel to the ends of the soil sample. Since this pad is slightly deviated from the calculated value, the strength parameters during such tests may slightly differ from the real figures.

The condition of soil strength under triaxial stress state (case *1b*) is traditionally determined by the Coulomb-Mohr theory of strength. In accordance with this theory, the diagram is constructed in the rectangular coordinate system $\tau_\alpha, \sigma_\alpha$, and the strength condition is estimated by the

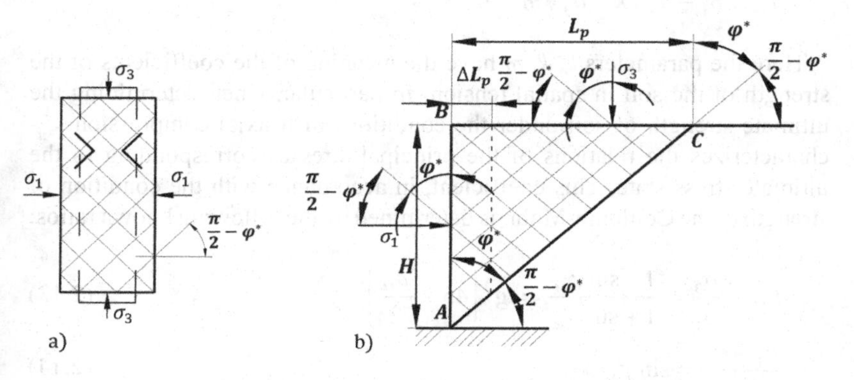

Figure 2.8 The state of stress and the situation changes in the passive surfaces.

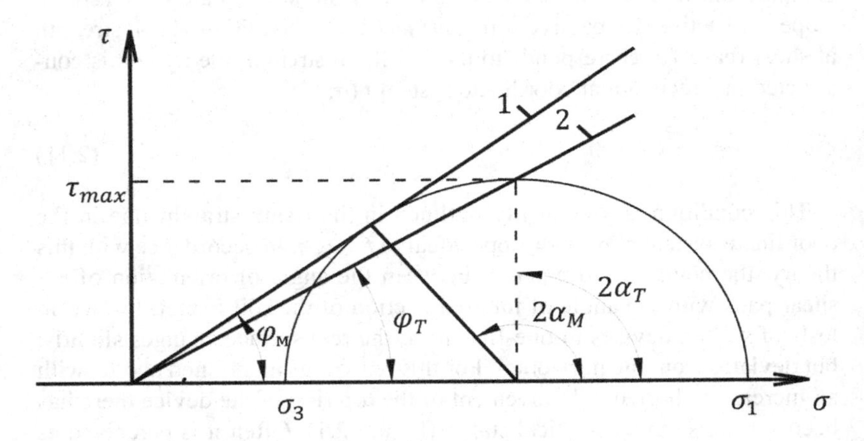

Figure 2.9 Graphical representation of the strength condition according to the Coulomb-Mohr theory -1 and Tresk Hill -2.

equation of the line drawn along the tangential line to the Mohr circles (Figure 2.9). In this case, the condition of the strength of soils under triaxial stress state is written in the form

$$f^* \equiv \sigma_3 - \sigma_1 * \xi = 0 \tag{2.12}$$

or

$$f^* \equiv \sigma_1 - \sigma_2 * k - \sigma_3 * m = 0$$

Here the parameters ξ, *k*, *m* have the meaning of the coefficients of the strength of the soil in spatial tension. In particular, when determining the ultimate strength of soil under the conditions of triaxial compression, ξ_u-characterizes the relations of the principal stresses corresponding to the ultimate stress state. This coefficient, in accordance with the condition of strength of the Coulomb-Mohr, is determined by the following known ratios:

$$\xi_M = \frac{\sigma_3}{\sigma_1} = \frac{1-\sin\varphi_M}{1+\sin\varphi_M} = tg^2\left(45 - \frac{\varphi_M}{2}\right) \tag{2.13}$$

$$\frac{\sigma_1 - \sigma_3}{\sigma_3 + \sigma_1} = \sin\varphi_M \tag{2.14}$$

As can be seen from equation (2.13), there is a rigid analytical relation between the principal stress ratios with the pad's deflection angle $\alpha_M = (\pi/4 \pm \varphi/2)$, with the coefficient of the internal friction angle φ_M, and with the lateral pressure coefficient ξ_M. However, as was shown above, in reality, the strength parameter φ is the angle of internal friction for soils, it is relatively constant and weakly depends on the type of stress state. Whereas the lateral pressure coefficient can vary over a wide range, depending on the type of stress state and soil movement. In addition to soil the actual orientation of the platforms shear differs from that calculated by the theory of Mohr-Coulomb $(\pi/4 \pm \varphi/2)$ and corresponds, as shown by our study (see section 1.3), the value of φ^* and φ^*и$(\pi/2 - \varphi^*)$, where $\varphi^* > \varphi$.

There are other criteria that evaluate the ultimate state of soils. For example, the strength criterion proposed by Tresk Hill [5]. In accordance with this theory, shear pads are deflected relative to the principal stresses by an angle of 45°(degrees). Then the strength condition will have the following form:

$$\frac{\sigma_3 - \sigma_1}{\sigma_3 + \sigma_1} = \operatorname{tg}\varphi_T$$

$$\xi = \frac{\sigma_3}{\sigma_1} = \frac{1 - tg\varphi_T}{1 + tg\varphi_T} \qquad (2.15)$$

where φ_T – friction angle of the theory of strength of Tresc Hill (Figure 2.9). In accordance with this theory of strength, the graph of the ultimate skirt line is constructed similarly to the Coulomb-Mohr theory of strength, except that the skirt of the ultimate line is drawn not as a tangent to the circle, but intersects it and connects to it at the point where $\tau_{max} = (\sigma_3 - \sigma_1)/2$. For this reason, the condition $\varphi_T < \varphi_M$ is always satisfied, and the angle of deviation of the shear pad is $\alpha_T = 45°$.

In accordance with the theory of Mises-Botkin, the condition of soil strength in space is depicted as a truncated cone with a central longitudinal axis tilted well to the three axes of the rectangular space (Figure 2.10). Graphically, the deviatorial stress is a perpendicular section made to the bisector, and the bisector itself characterizes the hydrostatic stress state. The deviatoric area is sometimes referred to as octahedral. The central axis is the direction cosines $l = n = m = 1/\sqrt{3}$. In the plane coordinate system of Mises strength theory, Botkin graphically describes a bisector still inclined to the axes of the principal stresses. A straight line with an angle of deviation by an angle $\theta = 45°$, where $\sigma_3 = \sigma_1$. The deviator axis is directed perpendicular to the bisector. The tangential stresses $\sigma_i = \sigma_1 - \sigma_3$ act on this site (Figure 2.9).

In the context of the spatial problem, the tensile strength of soil in the deviatoric plane is a circle. Taking into account the impact of the stress

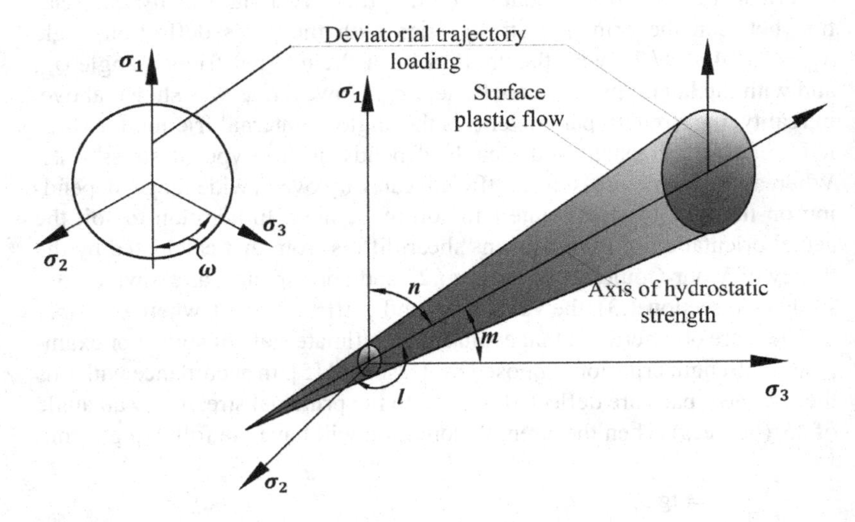

Figure 2.10 Mises-Botkin Chart conditions of strength.

state, it has the form of a convex triangle. G. Lomize and A. Kryzhanovskii obtained these results [8]. Our results showed that when a stress state $\mu = 1$ (triaxial tensile $\sigma_3 = \sigma_2 > \sigma_1$) and $\mu = 0$ (the strength of the soil torsional cylindrical samples of soil and one flatness shear) tensile strength and thus the angle of internal friction of soil coincide. The several underestimated results of the soil tensile strength were obtained during triaxial soil tests at $\mu = -1$ (triaxial crushing $\sigma_1 > \sigma_3 = \sigma_2$) (Table 1.3). The mathematical relationship between the angle ω and Lode-Nadai parameter μ has the form:

$$\mu = \frac{\sqrt{3}}{tg(\omega + \pi / 3)} \tag{2.16}$$

where $0 \le \omega \le \pi / 3$.

Based on the above we can conclude that the strength of the soil, as opposed to the continuous crystallization materials, has characteristics and is characterized primarily by the forces of friction between the contacts of mineral particles or aggregates (for clay soils). This greatly distorts the state of stress in the elementary volume of soil. For example, under the conditions of a plane task, the traditionally stressed state for crystallization continuous materials changes in accordance with the theory of Mohr. According to this theory, when the angle of deviation of the pad changes, the normal and tangential stresses change too. In this case, the area where only normal stresses act is called the principal one. If you try to subordinate the theory to the theory of Coulomb dry friction, it satisfies only one condition, when the orientation of the most dangerous area is $\alpha = \frac{\pi}{4} \mp \frac{\varphi}{2}$. However, as noted above, such a soil area for some other orientation and equal to φ^* and $\left(\frac{\pi}{2} - \varphi^*\right)$. Match with real soils is only possible if the condition $\varphi^* = 30°$. A study of the real properties of soils also shows that they differently resist totriaxial compression (crushing) and shear. In an analysis of the literature, numerous studies by various scientists suggest that this discrepancy may reach 10%–15%. Moreover the coefficient of lateral soil pressure is variable. Therefore, this amount related to the ultimate state of stress can also be taken as an additional condition of soil strength. Unfortunately, such specific properties are inherent only to soils, and the Coulomb-Mohr theory of strength does not take into account.

In accordance with the theories of Tresc(k)-Hill and Mises-Botkin we analyze the most dangerous sites tilted in respect to the axes of the main stresses at $\alpha = 45°$ (plane problem) and $\alpha = 54.8°$ (octahedral Playground in a three-dimensional problem). Under the conditions of a flat problem, the Mises – Botkin theory of soil strength graphically characterizes a straight line with an angle of deviation of $\theta = 45°$, where $\sigma_{2=3}$, and the deviator is perpendicular to it. Tangential stresses τ act on this pad. As can be seen from the analysis of the above theories, the orientation angle of the shear pads differs from the actual shear.

Comparison of soil strength conditions

As already noted above in practice, in solving flat tasks, the theory of strength of Coulomb, Coulomb-Mohr and Tresc(k)-Hill is used. To compare the results of these theories, for clarity, we consider a graphical interpretation of it.

1 Theory of Coulomb-Mohr strength. To build a strength diagram for this theory, we need to experimentally, on triaxial compression devices (stabilometer), determine the ratio of the main stresses corresponding to their limiting state, i.e. $\sigma_{1u} \geq \sigma_{(2=3u)}$. Suppose we take non-cohesive soils, such as sand, as an object of research. According to the test results, we will construct a Mohr's diagram and draw the limit envelope to it. The angle of deviation of this straight line from the abscissa axis σ is denoted by φ_M. Next, we define the remaining parameters of the condition corresponding to the theory of Coulomb-Mohr strength: $\alpha / 2 = (\pi / 4 \pm \varphi / 2)$ is the deflection angle of the shear area; $\xi_u = \sigma_3 / \sigma_1$ is a coefficient characterizing the ratio of the smallest and largest principal

stresses under the conditions of the limiting stress state. Define the main design parameters:

$$\sigma_m = (\sigma_1 + \sigma_3) / 2; R = (\sigma_1 - \sigma_3) / 2; \sin\varphi_M = R/\sigma_m$$

from where based on calculations

$$cos\theta = \frac{\tau_\alpha}{R};\ sin\theta = \frac{\sigma_m - \sigma_\propto}{R};$$

$$\tau_\alpha = R * cos\theta \text{ и } \sigma_\alpha = \sigma_m - R * sin\theta.$$

2 The theory of the strength of Tresc(k)-Hill. To build a strength diagram for this theory, we need to experimentally, also on devices of three-axis compression (stabilometer), determine the ratio of the main stresses corresponding to their ultimate state, i.e. $\sigma_{1u} \geq \sigma_{(2=3u)}$. At the same time, on the strength diagram, we will connect the origin of coordinates O with a line M, where the condition $\tau_\alpha = \tau_{max} = R$ (Figure 2.9, line –2) is fulfilled:

$$tg\varphi_T = \frac{\tau_\alpha}{\sigma_\propto} = \frac{R * cos\beta}{\sigma_m - R * sin\beta};$$

Under condition (2.13), the theory of Coulomb-Mohr strength is valid.

$$tg\varphi_M = \frac{\tau_\alpha}{\sigma_\propto} = \frac{R * cos\varphi_M}{\sigma_m - R * sin\varphi_M}$$

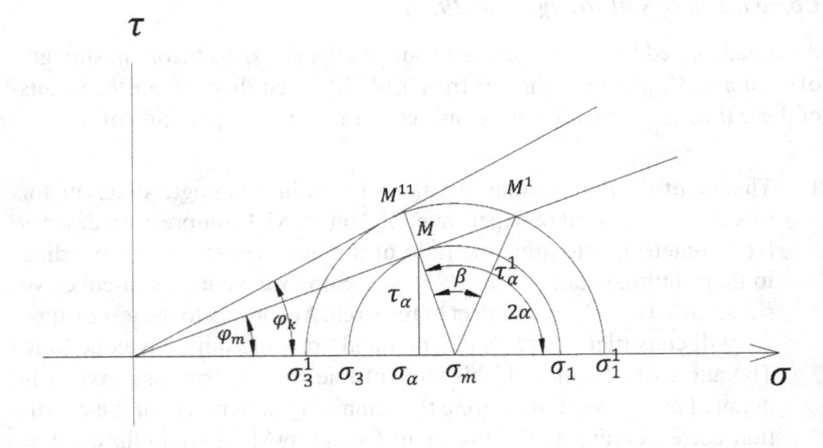

Figure 2.11 Comparison of the strength condition of the soil and its graphic application.

When the condition $\tau_\alpha = \tau_{amax} = R$ is fulfilled, the condition corresponding to the Tresc(k)-Hill strength theory holds.

$$tg\varphi_T = \frac{\tau_{\alpha max}}{\sigma_\propto} = \frac{R}{\sigma_m}$$

In this case, the envelope line will shift from point M to the point corresponding to the peak of the circle. In this case, the condition $\varphi_T < \varphi_M$; [[$\tau_\alpha = \tau_{amax}$ and the angle of rotation of the shear area $\frac{\alpha}{2} = 45°$ is satisfied. Then we obtain the value of the angle of internal friction of the corresponding Coulomb strength theory; i.e. on the strength diagram, this corresponds to the point M^{11}.

$$tg\varphi_T < tg\varphi_K = \frac{\tau_\alpha}{\sigma_\propto}$$

When this condition is satisfied, $\varphi_K > \varphi_M > \varphi_T$; $\tau_\alpha = \tau_{\alpha max}$ and angle of rotation $\frac{\alpha}{2} = 45°$.

3 Theory of Coulomb-Mohr conditions under pure shear. To build a strength diagram for this theory, we use the results of experiments conducted on devices of triaxial compression with torsion. In this case, the ratio of main stresses is experimentally determined corresponding to their limiting state, i.e. τ_α, σ_m

$$tg\varphi_K = \frac{\tau_\alpha}{\sigma_m};\ \sigma_m = \frac{\sigma_1 + \sigma_3}{2}$$

In this case, the principal stresses $\sigma_1 \geq \sigma_{2=3}$ are already determined by calculation. In accordance with the Mohr's circle.

$$\sigma_1 = \tau_\alpha\left(\frac{1}{cos\varphi_K} + \frac{2}{sin2\varphi_K}\right);\ \sigma_3 = \sigma_1 - \frac{2\tau_\alpha}{cos\varphi_K}$$

$$\sigma_m = \frac{\sigma_1 + \sigma_3}{2} = \frac{\sigma_1 + \tau_\alpha\left(\frac{1}{cos\varphi_K} + \frac{2}{sin2\varphi_K}\right) - \frac{2\tau_\alpha}{cos\varphi_K}}{2};$$

Based on this, it is possible to draw an important conclusion that when calculating the ultimate stress state of a base from the action of critical forces, it is necessary to take into account three types of strength parameters: Coulomb, Coulomb-Mohr, and Tresc(k)-Hill, i.e. φ_K; φ_M, ξ_u and φ_T, ξ_u. Moreover, for them the condition $\varphi_K > \varphi_M > \varphi_T$ is satisfied. Unlike the first model in (2.11) and (2.12), it is possible, depending on the magnitude of the main stresses $\sigma_1 > \sigma_3$, to determine the component stresses τ_θ, σ_θ acting on any site

($0 \leq \theta \leq \pi/2$), as well as the deviation angle of the shear area $\frac{\alpha_M}{2} = (\pi/4 \pm \varphi/2)$, $\frac{\alpha_T}{2} = 45°$. The results of numerous experiments, and their comparison with the obtained analytical calculations, showed the presence of differences between them. In particular, it has been established that $\varphi_K > \varphi_M > \varphi_T$, and that the deviation angle of the shear area is $45° \leq \frac{\alpha}{2} \leq (\pi/4 \pm \varphi/2)$. Such a discrepancy naturally affects the final results of calculations of the strength and stability of the bases of buildings and structures. Therefore, we have attempted to develop a theory of strength that allows us to equally satisfy both the theory of the strength of Coulomb (2.11) and Coulomb-Mohr (2.12). The results of such studies are given in section 2.3.

Consideration of the bonding strength of clay soils

Above, the strength condition was considered mainly for incoherent soils. Below we show the validity of these fundamental equations as applied to coherent soils.

In the beginning, we consider the theory of Coulomb strength as applied to clayey soils.

$$\tau_\alpha = \sigma_\alpha tg\varphi_K + c = \left(\sigma_1 + \frac{c}{tg\varphi_K}\right) tg\varphi_K = (\sigma_1 + P_e) tg\varphi_K = \sigma_{ЭКВ} tg\varphi_K$$

Where $P_e = \frac{A}{tg\varphi_K}$ – bonded force characterizes bulk negative pressures equal in magnitude to the resistance of soils to comprehensive stretching. Applied to the theory of Coulomb Mohr strength:

$$\sigma_{1ЭКВ} = \sigma_1 + P_e \text{ и } \sigma_{3ЭКВ} = \sigma_3 + P_e$$

$$\sigma_m = \frac{\sigma_1 + \sigma_3 + 2P_e}{2}; R = \frac{\sigma_1 - \sigma_3}{2};$$

$$sin\varphi_m = \frac{R}{\sigma_m} = \frac{\sigma_1 - \sigma_3}{\sigma_1 + \sigma_3 + 2P_e} = \frac{\sigma_1 - \sigma_3}{\sigma_{ЭКВ1} + \sigma_{3ЭКВ}}$$

Below is the model proposed by the authors, which makes it possible to describe the stress state at a point and to search for the most dangerous shear site. This model to one degree or another takes into account the more realistic properties of soils. When developing this model, it is taken into account that the normal stress at this point is a set of limiting proportions of normal stresses sufficient to shift the moving mineral particles along the most dangerous area, the deviation from the angle φ^* and (π / 2 – φ^*), where $\varphi^* > \varphi$. $\xi_u = \frac{\sigma_3}{\sigma_1}$ and allows you to determine the position of real shear surfaces. It should not rely on stress analysis at a point.

2.3. Alternative concepts of the theory of strength of sandy soils

The ratio of reactive R_σ *and internal stresses* $\sigma_\theta, \tau_\theta$ *obeys the law circle and ground shear denied in relation to the horizon at an angle* $\theta^* = \left[\frac{\pi}{2} - (\theta + \alpha_{\xi a})\right]$ *in the passive and* $\theta^* = (\theta + \alpha_{\xi p})$ *with the active loading.*

Below, we analyze an alternative version of the theory of strength of soils, which will take as a basis the following postulates: the angle of internal friction and the coefficient of lateral earth pressure are taken as independent parameters of the equation of strength; orientation playgrounds shear with respect to the direction of the principal stresses is assumed to be:

with the active and passive loadings - $\theta^* = \left[\frac{\pi}{2} - (\theta + \alpha_{\xi a})\right]$, $\theta^* = (\theta + \alpha_{\xi a})$ or $\varphi^* = \left[\frac{\pi}{2} - (\varphi + \alpha_{\xi a})\right]$, $\varphi^* = (\varphi + \alpha_{\xi a})$ when $\theta = \varphi$ (Figure 2.7, 2.8).

To solve this problem, we consider a rectangular coordinate system $f = (\sigma_3, \sigma_1)$. The vertical axis establishes normal voltage σ_1 and assumes that its direction corresponds to the direction of gravity. The horizontal axis will set the direction σ_3. The angle of orientation of the coordinate system $f = (\sigma_3, \sigma_1)$ may vary depending on the outline terms of the problem being solved. Depending on the ratio of the limiting stress σ_1 and σ_3 problems can be of three types:

1 $\sigma_1 > \sigma_3$ resistance ground (see task 1);
2 $\sigma_3 = \sigma_1$ hydrostatic compression;
3 $\sigma_3 > \sigma_1$ passive resistance of the soil (see task 3).

In order to avoid confusion in choosing a global coordinate system, we take the direction of the main stresses σ_1 strictly vertically (Fig. 2.11).

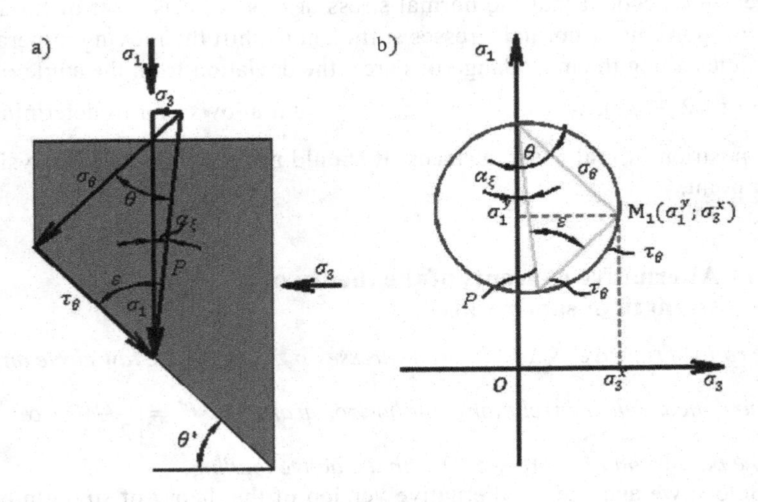

Figure 2.11

Task 1: Active soil pressure

Let's put on this graph the limiting vertical σ_1 and horizontal σ_3 stresses (Figure 2.11). Moreover, in this case, the direction of the axis σ_1 coincides with the axis of gravity. Graphically we connect two points σ_1, σ_3 on this plane, we get a regular triangle. The ratio of these voltages on the graph characterizes the lateral pressure coefficient. The direction of the full voltage vector P is oriented towards the site where the smallest principal stress (1.25) (Figure 2.13) operates. The angle of deviation of the full vector of voltage P from the greatest stress σ_1 is $\alpha\xi$. In relation to our example, we denote this angle by $\alpha\xi$.

$$\xi = tg\alpha_\xi = \frac{\sigma_3}{\sigma_1} \tag{2.17}$$

In accordance with the theory of Coulomb-Mohr strength, the magnitude of the total stress vector P is determined by the equation

$$P = \sqrt{\sigma_\theta^2 + \tau_\theta^2} = \frac{\sigma_1 + \sigma_3}{2}\sqrt{1 - \left(\frac{\sigma_1 - \sigma_3}{\sigma_1 - \sigma_3}\right)^2} \tag{2.18}$$

This equation can be written taking into account the cosines:

$$P = \sqrt{\sigma_1^2 l^2 + \sigma_3^2 s^2} \tag{2.19}$$

where $l = cos\theta^*$, $s = cos\left(\frac{\pi}{2} - \theta^*\right)$ cosine guides.

In accordance with the diagram (Figure 2.13), the magnitude of the hollow stress vector corresponding to the Coulomb strength equation can be written in the following form:

$$P^{*} = \frac{P}{ctg\left(\varphi_{K} - \varphi_{M}\right)} \tag{2.20}$$

where φ_{κ}, φ_{M} – respectively, the angle of internal friction determined on the shear devices and the τ, σ diagram and the Mohr diagram from the results of the triaxial tests.

We draw a circle with the diameter D = P* in the direction of the full stress vector P* at the end of the pole σ_1. Next, Axis σ_1 draw a line with a deviation angle $\theta^* = (\theta + \alpha_\xi)$ to the intersection point M (σ_1; σ_3) lying on the surface of the circle. With this, the two lines form a right-angled triangle. Thus, the left side of the vector P* characterizes the main external, and the right side the internal voltages at the site, deviated from the horizontal by the angle $\theta^* = (\theta + \alpha_\xi)$. The length of this straight segment (line σ_1; M (σ_1; σ_3) is equal to normal σ_θ, and the opposite side of the triangle is equal to the value of tangential stresses τ_θ acting on this site. The ratio of these stresses depends on the angle of deviation θ. The angle of deflection of the vector of stresses τ_θ, and σ_θ in relation to the axes σ_3 and σ_1, is $(\theta + \alpha_\xi)$. The presented diagram, unlike Mohr's theory of strength, characterizes the limiting state conditions satisfying both equations (2.11) and (2.12). Using the known value of the total stress vector P, taking into account equation (1.25), we determine the composite stresses acting along the inclined pad

a the limiting stresses at the shear site σ_θ, τ_θ and P* are known

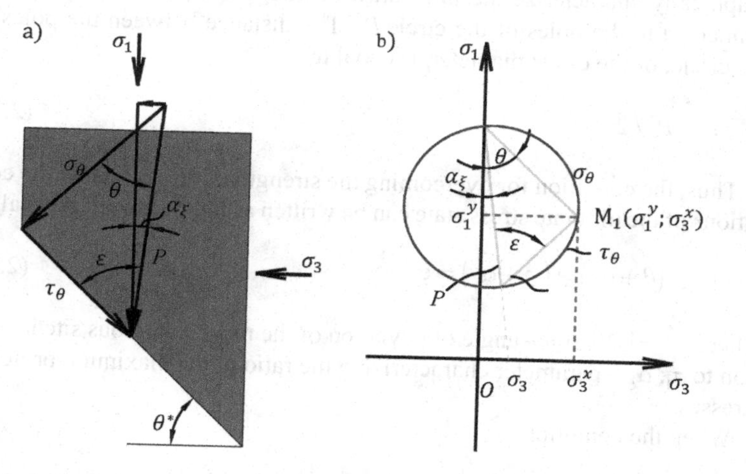

Figure 2.12 Graphic interpretation of the soil strength condition by the author: a) active loading in the plane stress state, b) soil strength diagram.

$$\sigma_1 = \frac{P^*}{\sqrt{l^2 + \xi^2 s^2}} \tag{2.21}$$

$$\sigma_3 = \frac{P^*}{\sqrt{l^2 / \xi^2 + s^2}} \tag{2.22}$$

b *the limiting principal stresses are known.* σ_1, σ_3 and P^*

$$\sigma_\theta = P^* \cos\theta \tag{2.23}$$

$$\tau_\theta = P^* \sin\theta \tag{2.24}$$

The coordinates of the point $M(\sigma_1^y; \sigma_3^x)$, *t, the intersection of two straight lines* σ_θ и τ_θ *on the surface are determined by the following equations:*

$$\sigma_3^x = \sigma_\theta \sin(\theta + \alpha_\xi)$$

$$\sigma_1^y = \sigma_1 - P\cos(\theta + \alpha_\xi)\cos(\theta) \tag{2.25}$$

where – $\theta^* = (\theta + \alpha_\xi) - \left(\frac{\pi}{2} - \theta^*\right)$ the angle of deviation of the displacement area relative to the areas where the main stresses act $\sigma_1; \sigma_3$

When $\theta = \varphi$, and $\theta^* = (\varphi + \alpha_\xi)$ condition $tg\varphi = \frac{\tau_\theta}{\sigma_\theta}$

The vector connecting normal stresses σ_θ and tangential stresses τ_θ, is the reactive resultant of forces P acting on an inclined pad.

In the equations (2.20, 2.21) σ_θ normal and tangential stresses τ_θ graphically characterize the lines drawn through the point $M(\sigma_3^x, \sigma_1^y)$ and connected to the poles of the circle P^*. The distance between the poles of the circle, or the circle diameter, is equal to

$$R = P^* / 2 \tag{2.26}$$

Thus, the condition for overcoming the strength of the soil under the conditions of the ultimate stress state can be written as the following inequality:

$$\theta^* = (\theta + \alpha_\xi) \geq (\varphi + \alpha_\xi) = \varphi^* \tag{2.27}$$

where $\varphi^* = (\varphi + \alpha_\xi)$ – angle of deviation of the most dangerous site in relation to σ_1; α_ξ – parameter characterizing the ratio of the maximum principal stresses.

When the condition

$$\theta^* = (\theta + \alpha_\xi) < (\varphi + \alpha_\xi) = \varphi^* \tag{2.28}$$

the ground will always be in the state to limit, or a shear in these areas is not possible.

Based on the foregoing, an important conclusion can be drawn that, in order for a shift to occur at the most dangerous pads, the fulfillment of the condition $\theta \geq \varphi$ is insufficient. A second prerequisite for shearing the soil at this location is the other conditions $\frac{\sigma_3}{\sigma_1} = \xi \leq \xi_u$. Targeting the most dangerous area in relation to the site, where there are the greatest principal stress σ_1 equal to $\varphi^* = (\varphi + \alpha_\xi)$.

Let's analyze what was said on the soil strength diagram (Figure 2.12). Suppose that we know the limiting ratio of the stress state of the soil $\frac{\sigma_3}{\sigma_1} = \xi_u$. According to these stresses in the diagram, we determine the total stress vector P^* acting on an inclined platform. To this at an angle φ draw straight lines until they intersect with a circle whose radius is half the length of the vector P^*. On it we define the limit values of normal σ_θ and tangents τ_θ stresses acting on this site.

For comparison, we consider the same problem in the Mohr voltage diagram. In this case, the magnitude of the total stress vector P defined as a straight line segment OM operating on a site rejected at an angle $45° \leq (\varphi + \alpha_\xi) \leq (\pi / 4 \pm \varphi / 2)$. In this case, an angle forms on the Mohr's diagram of stresses. $\theta_{M\theta} \leq \varphi_M$. Next, taking the value of the total stress vector P as the radius (observing the fulfillment of the basic law of mechanics, i.e. the equilibrium condition), we rotate it before reaching the angle $\theta = \varphi_k$. At the same time, we will have to go beyond the Mohr's strength diagram and combine with the Coulomb strength diagram.

Further for this site corresponding $\frac{\alpha}{2} = (\varphi + \alpha_\xi)$, as shown in the diagram, we define composite normal σ_θ and shear τ_θ stresses. Figure 2.13 presents the theory of strength proposed by the authors, built on the Mohr stress diagram.

From the diagrams presented above, it follows that within the limits of the deformability of the soil, when the condition $\frac{\sigma_3}{\sigma_1} > \xi_u$ and $\theta^* < \varphi^* = (\varphi + \alpha_\xi)$ the strength at a point can be determined with sufficient accuracy by the theory of stresses proposed by Mohr. As the stress state approaches its critical point, shear surfaces are formed with a deflection angle approaching $\theta^* = \varphi^* = (\varphi + \alpha_\xi)$. In this position in the ground tangential stress τ_θ increases at a faster pace and normal σ_θ is slightly reduced. For this reason, in soils under conditions of a limiting stress state, especially in sands, an expansion is observed (loosening, see Figure 1.16). From the presented two methods for calculating the limiting states (equations 2.21–2.22 and 2.23–2.24), it follows that they satisfactorily coincide and can be used to calculate the strength and stability of the soil foundation. Changes in the composite stresses

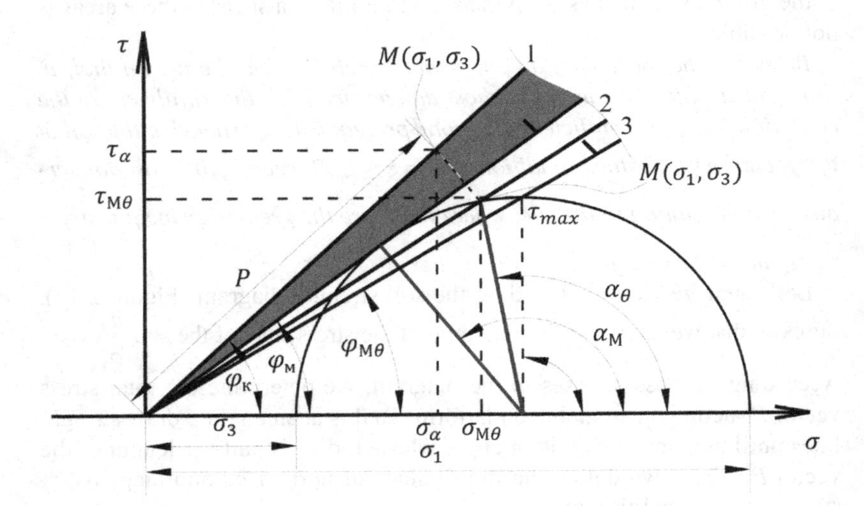

Figure 2.13 Graphic representations of the strength condition according to the Coulomb-Mohr theory (1) and Treska-Hill (2).

$\sigma_\theta, \tau_\theta$ at the ultimate shear site can also be traced by the Mohr diagram (Figure 2.13). In this case, starting from point 0-$M(\sigma_1, \sigma_3)$, *the circle lying on the surface of the ultimate skirt line of the straight line to the Mohr circle is drawn by* $R = P$ *until it is aligned with the ultimate skirt line of Coulomb. Yes, there will be a slight increase* $P^* > P$, *stress* τ_θ *and a slight decrease in* σ_θ. *The angle will increase accordingly.* $\varphi_{M \rightarrow} \varphi_K$. *Based on the above, we can conclude that for describing the stress state at a point for soils that are in the pre-limit state, to solve practical engineering geotechnical problems, one can use the Coulomb-Mohr stress theory. Moreover, all the parameters corresponding to this theory will be related to the deformation ones. In the ultimate stress state, the Coulomb-Mohr strength theory gives inaccurate results of the main parameters of strength: the overestimated results of the deviation angle of the shear area and the underestimated results of the angle of internal friction. In addition to this, at the shear site, under the conditions of the limiting stress state, a transformation occurs, i.e. a slight decrease in normal and a slight increase in tangential stresses. It is these main disadvantages of the Coulomb-Mohr strength theory that are taken into account in the proposed theory of the strength of soils.*

For the platform with the guide cosines and the angle of deflection $\theta = \varphi = 39° \varphi^* = (\varphi + \alpha_\xi) = 39 + 15 = 1 = 54°$ *и* $s = 36°$. Let us compare the obtained results of experiments and compare them with the results of

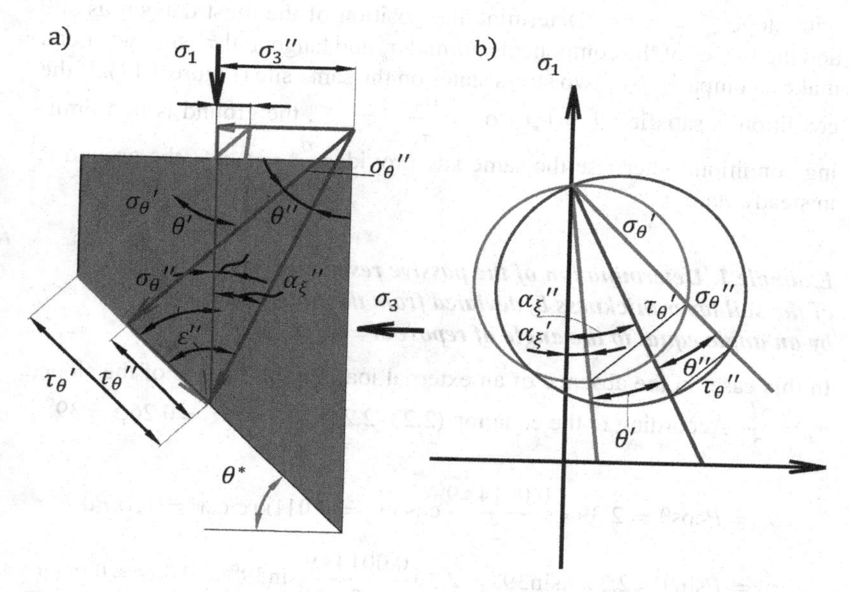

Figure 2.14 Stress comparison $\sigma_\theta, \tau_\theta$ on a fixed platform deviated from the horizontal by an angle θ^*, at different ratios of main stress σ_1, σ_3.

Table 2.1 Results

Theories	σ_1	σ_3	ξ_u	α_ξ	l	s	P	σ_θ	τ_θ	θ
	kPa						*kPa*			*Rad/(grade)*
2.13 2.14	760	200	0.26	15	0.588	0.81	475	393	266	0.67/(34)
2.21- 2.24	760	200	0.26	15	0.588	0.81	475	369	299	0.81/(39)
difference								−24	+33	+0.14/(5)

the classical theory of the determination of Mohr's stresses. Table 2.1 presents these results.

Let us compare the proposed theory in the ultimate condition and before the ultimate condition. To do this, we plot two types of stress states on the diagram. These stresses will be applied to the soil strength diagram. Determine the magnitude and direction of the resultant vector P^{11} and at its poles draw a circle. Further, on the same graph, for comparison, we plot the contours of the limiting stresses. For this, from the point of characterizing the greatest principal stress, for example, σ_1, draw a line with a limit angle equal to α_ξ^1 or with slope $\xi_u = \sigma_3 / \sigma_1$. Determine the position of the most dangerous site and the values of the components normal σ_θ^1 and tangential τ_θ^1 stresses. Let's make a comparison of two stress states on the same site (Figure 2.14). If the condition is satisfied $\theta^* = (\varphi + \alpha_\xi^1)$ и $\frac{\sigma_3}{\sigma_1} = \xi \le \xi_u$, the ground is in a limiting condition, otherwise the same site provided $\frac{\sigma_3}{\sigma_1} = \xi > \xi_u$ the ground is in steady state.

Example 1. Determination of the passive resistance of the soil layer thickness b, deviated from the horizontal by an angle equal to the angle of repose $\beta = \varphi, \sigma_1 > \sigma_3$

In this case, in the absence of an external load on the surface of the slope, $\sigma_3 = \frac{\gamma b}{2}$. According to the equation (2.23–2.24), $\alpha_\xi = 15\xi = 0.26\varphi = 39°$

$$\sigma_\theta = P\cos\theta = 2.39 * \frac{0.0014 * 9}{2}\cos 39° = 0.011 \text{кгс/см}^2 = 1.1 \kappa Pa$$

$$\tau_\theta = P\sin\theta = 2.39\sigma_3 \sin 39° = 2.39\frac{0.0014 * 9}{2}\sin 39° = 0.009 = 0.9 \text{кPa}$$

$$P = \sqrt{\sigma_1^2 l^2 + \sigma_3^2 s^2} = \sigma_3\sqrt{\frac{1}{\xi^2}l^2 + s^2} = \sigma_3\sqrt{\frac{1}{0.067}0.34 + 0.66} = 2.39\sigma_3$$

$$l = cos(\varphi + \alpha_\xi^1) = cos(39 + 15) = 0.59$$

The total force acting on the side surface by means of a stamp area A = 9 × 9 = 81 см^2

$$\text{N} = \sigma_1 * A = 0.024 * 81 = 2\kappa\textit{гс} = 20H$$

This implies that the theory of strength of Mohr gives overestimated results values for the angle of internal friction and the deflection angle of the shear area. According to the proposed method of calculation, the parameters of strength and coefficient of lateral pressure are given separately, and for this reason, these design parameters are close to the actual experimental results obtained. The parameter coefficient of lateral pressure in the passive and active loadings is variable and depends on the type of the stress state and the value of the lateral movement of soil. For example, when tested sands coefficient of lateral pressure, at rest, $\varepsilon = 0$ is $\xi \cong 0.17$, at the same time in the presence of lateral movement in the trough, respectively, and tests with the active stabilometric pressure $\xi = 0.08 / 0.26$. However, the passive

Table 2.2 Comparison of strength parameters and deviation angles platforms shear

Name of the experimental setup and the type of the stress state	*Max and min ultimate stresses (experiment)*			*According to the Mohr-Coulomb model*			*According to the proposed model*			
	σ_3 *кPa*	σ_1 *кPa*	$\xi = \frac{\sigma_3}{\sigma_1}$	σ_φ *кPa*	τ_φ *кPa*	φ_M *(°)*	σ_φ *кPa*	τ_φ *кPa*	φ *(°)*	φ^* *(°)*
1 – flat tray, passive pressure, horizontal position of tray	9.2	88.57	0.104	16.7	23.6	54	40.9	33	39	45
2 – a flat tray, a passive pressure tilt tray (Example 3)	8	1	0.12	1.8	2.2	51	6.26	5.07	39	46
3 – a flat tray, active pressure horizontal tray	18.5	230	0.08	34.2	55.5	58	105.7	85	39	44
4 – triaxial type A (crush)	200	760	0.263	316	227	35.6	369	299	39	54
5 – stabilometry of type B (tension)	104	460	0.23	170	137	39	220	178	39	52
	104	480	0.22	170	143	40	228	185	39	51.2
6 – The device torsion tubular samples. Hydrostatic with torsion (2.10)	31(24)	104(104)	0.23	50	40	38	50	40	39	52
	70(48)	270(209)	0.23	100	81	38	100	81	39	52
	140(95)	538(410)	0.23	200	161	38	200	161	39	52

Note. $\sigma_{1,3}$ *applied to soil tests on the instruments torsion: outside of the brackets the calculations are made using the Mohr model, and inside the brackets the calculations use the model proposed by the authors.*

pressure $\xi = 0.11 / 0.23$ (see Table 2.1). The question of having a functional relationship between the parameters of strength and the coefficient of lateral earth pressure will be discussed below.

Example 2. Determination of the active resistance of the soil on the results of soil testing for triaxial compression with torsion

The results of the soil strength test are given in Table 1.3. In particular, $\varphi = 39°$; $\alpha_\xi = 13°$; $\sigma_\theta = 200\kappa Pa$ *u* $\tau_\theta = 161\kappa Pa$. According to the equation (1.25) we determine the component of the total stress vector on an inclined platform

$$P = \sqrt{\sigma_1^2 l^2 + \sigma_3^2 s^2} = \sqrt{\sigma_\theta^2 + \tau_\theta^2} = \sqrt{200^2 + 161^2} = 257\kappa Pa$$

$$P = \sigma_1 \sqrt{l^2 + \xi^2 s^2}$$

$$l = \cos(39 + 13) = \cos 52° = 0.61 \text{ and } s = \cos(90 - 52) = \cos 38° = 0.79$$

$$\sigma_1 = \frac{P}{\sqrt{l^2 + \xi^2 s^2}} = \frac{257}{\sqrt{0.37 + 0.053 * 0.63}} = 410\ \kappa Pa$$

$$\sigma_3 = \xi \sigma_1 = 0.23 * 410 = 95\ \kappa Pa$$

For comparison, the same values will be determined in accordance with Mohr's theory for $\varphi = 35.6°$, derived from triaxial tests

$$\sigma_m = \frac{\tau_\theta}{cos\varphi sin\varphi} = \frac{2\tau_\theta}{sin2\varphi} \tag{2.29}$$

$$\sigma_m = \frac{2 * 161}{\sin(2 * 35.6)} = 340\kappa Pa$$

From the equations:

$$\sigma_1 + \sigma_3 = \frac{4\tau_\theta}{sin2\varphi} \text{ и } \sigma_1 - \sigma_3 = \frac{2\tau_\theta}{cos\varphi} \tag{2.30}$$

Define

$$\sigma_1 = \tau_\theta \left(\frac{1}{cos\varphi} + \frac{2}{sin2\varphi} \right) \tag{2.31}$$

$$\sigma_1 = 161 \left(\frac{1}{0.81} + \frac{2}{0.95} \right) = 538\kappa Pa$$

$$\sigma_3 = \sigma_1 - \frac{2\tau_\theta}{cos\varphi} \tag{2.32}$$

$$\sigma_3 = 538 - \frac{2*161}{0.81} = 140\kappa Pa$$

This example also shows the influence of the deviation angle of the shear area and the lateral pressure coefficient on the stressed state.

2.4. Theory of strength of cohesive soils and its practical application

Section 2.1 examined the physical conditions of the concept of engagement of granular soil. It was pointed out that to solve the practical problems of all the components of the strength located inside the bracketed equation (2.7) can be divided into external, gravity and volume (for example, power connectivity, the hydrodynamic pressure of groundwater). If the external and gravitational force can be considered as the value of the three-dimensional vector, which includes the internal forces of cohesion, they are scalar. Natural connectivity for clay soils depends on the history and geological conditions of the formation of sedimentary rocks. Without going into the physical and chemical nature of the concept of connectivity of clay soils, we note that their strength depends on the ratio of clay content (clay particles of less than 0.005 mm in size) and dust particles (dust particles of 0.05–0.005 mm in size) as well as on the chemical composition of salts contained in the soil. For this reason, their strength is strongly influenced by soil moisture and density.

Based on these properties, all clay soils are conventionally divided into sandy loam, loam and clay. Loam is mainly composed of dust particles (clay mineral content of at least 10% by weight). The mechanical properties are very similar to those of sandy soils and have low strength of connection. For clay soils include sedimentary rocks composed predominantly of the mineral kaolinite and montmorillonite in excess of the weight by more than 30%. Therefore in addition to their plastic properties, these soils also show rheological (viscous) properties. With increasing soil moisture the last appear stronger. Loam is an intermediary between the sandy loams and clays. Clay particles in this category primer are from about 10% to 30% by weight. Depending on the content of clay particles, clay loams are divided into light, medium and heavy. Heavy loam in its mechanical properties is close to clay.

The main mechanical parameter of strength for these soils is the angle of internal friction and cohesion. Parameters of the angle of internal friction and cohesion for clay soils cannot be identified as being similar to the parameters characteristic of granular non-cohesive soils. For clay soils, they have a more complex system of interaction, so most likely they can be arbitrarily taken as mathematical equation parameters of safety. For this reason, the physical strength of the essence for clayey soils largely depends on their condition (index hardness or consistency). As part of this book, we will consider clay soils having plastic properties, or clay soils with higher hardness plastic consistency.

Because of the strength of the proposed diagram (Figure 2.15) that

$$\Delta P = c / sin\theta) \tag{2.33}$$

$$\Delta\sigma_{1c} = \Delta P * cos\alpha_{\xi} = c\frac{cos\alpha_{\xi}}{sin\theta} \tag{2.34}$$

$$\Delta\sigma_{3c} = \Delta P * sin\alpha_{\xi} = c\frac{sin\alpha_{\xi}}{sin\theta} \tag{2.35}$$

$$\sigma_{1c} = \sigma_{1\varphi} + \Delta\sigma_{1c} \tag{2.36}$$

$$\sigma_{3c} = \sigma_{3\varphi} + \Delta\sigma_{3c} \tag{2.37}$$

The strength diagram will be presented in the form of two coordinate systems $f_1(\sigma_3, \sigma_1)$ and $f_2(\sigma_{3c}, \sigma_{1c})$. The first characterizes the strength condition

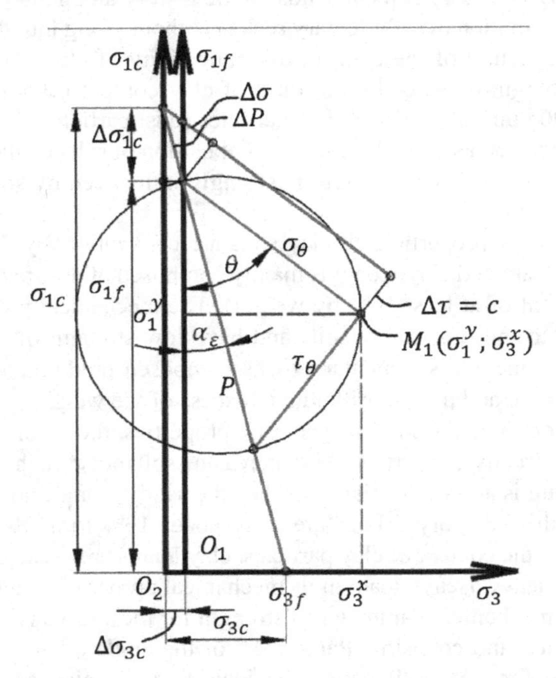

Figure 2.15 Graphic interpretation of the cohesive soils strength condition.

of soils, where only the influence of the parameter of the angle of internal friction is taken into account. To take into account the connectedness of the soil we will increment the vector P ΔP. Moreover, the initial condition must be met $\Delta P = c / sin\theta$. The increment values should correspond to equation (4.1). Origin of coordinates O_1 $f(\sigma_1, \sigma_3)$ move along the line OM_1 on O_2 $f_1(\sigma_1', \sigma_3')$. At the same time, another coordinate system is formed with the origin at the point O_2.

When the components of the voltage, acting on the site, deviated from the horizontal angle θ, accordingly will be equal

$$\Delta\sigma_{\theta c} = c\frac{cos(\theta)}{sin(\theta)} = P_5 \tag{2.38}$$

$$\Delta\tau_{\theta c} = c \tag{2.39}$$

for $\alpha_\xi = 0$, i.e. uniaxial compression case $\Delta\tau_{\theta c} = c$.

If the limiting principal stresses obtained for clay soils with a natural structure are known, then they can be used to analytically determine the strength of non-structured soils. Taking into account (2.36 and 2.37) we get

$$\sigma_{1\varphi} = \sigma_{1c} - \Delta\sigma_{1c} \tag{2.40}$$

$$\sigma_{3\varphi} = \sigma_{3c} - \Delta\sigma_{3c} \tag{2.41}$$

Equations (2.52) and (2.53) allow solving the problems of the theory of the strength of soils, if the stress state at a point is known.

Example. Determination of the cohesive soil strength parameter from the results of uniaxial tests (Figure 2.16)

Assume that we know

$$\alpha_\xi = 0; \sigma_{1f} = 0; \sigma_{1c} = 90 кPa \text{ и } \varphi = 22°$$

$$\Delta P = c / sin\theta (2.47)$$

$$\Delta\sigma_{1c} = \Delta P * cos\alpha_\xi$$

$$\Delta\sigma_{3c} = \Delta P * sin\alpha_\xi =$$

$$\sigma_{1c} = \sigma_{1\varphi} + \Delta\sigma_{1c} = \sigma_{1\varphi} + \Delta P * cos\alpha_\xi = \sigma_{1\varphi} + (c / sin\theta) * cos\alpha_\xi$$

$$\sigma_{1c} = 0 + (c / 0.37) * 1 = 90$$

$$c = 90 * 0.37 = 33.3 \text{ кPa}$$

$$\sigma_{3c} = \sigma_{3\varphi} + \Delta\sigma_{3c} = 0 + 90 * 0.37 = 33.3 \text{ кPa}$$

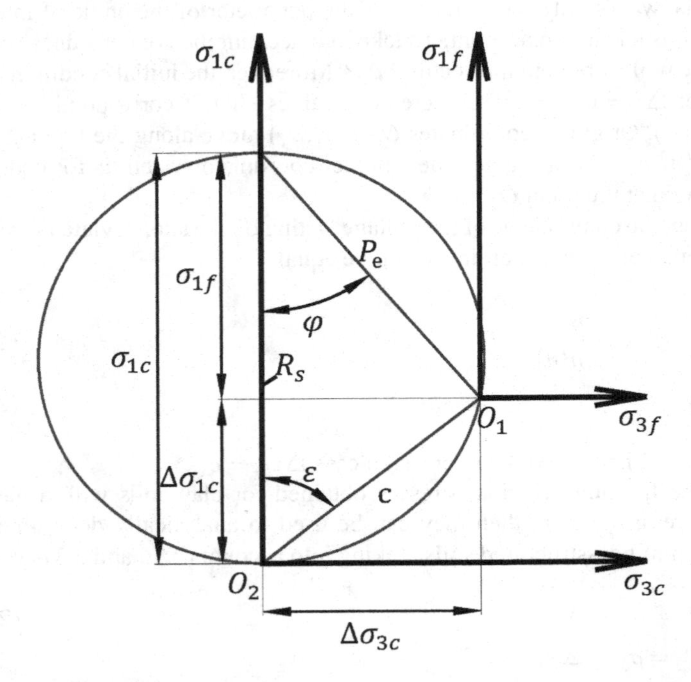

Figure 2.16 Graphic interpretation of the strength condition of cohesive soils under uniaxial compression.

$$\Delta P = \Delta\sigma_{1c} = \frac{c}{sin\theta} = \frac{33.3}{0.37} = 90\text{кPa}$$

2.4.1. *Determination of composite stresses at an arbitrary site deviating from the direction of the main stresses*

Task 1: Resistance of clay soils with active pressure.

$$\tau_\theta = P_f sin\theta + P_c = P_f sin\theta + c$$

$$\sigma_\theta = P_f cos\theta + \frac{c}{tg\theta}$$

$$\sigma_1 = \frac{P}{\sqrt{l^2 + \xi^2 s^2}} = \frac{\sqrt{\sigma_\theta^2 + \tau_\theta^2}}{\sqrt{l^2 + \xi^2 s^2}} \tag{2.42}$$

$$\sigma_3 = \xi\sigma_1 \tag{2.43}$$

Moistened clay soils prior to fracture undergo significant volumetric and deviatoric strain. This circumstance is the main cause of damage, and sometimes destruction of buildings and structures before the limit states subgrade. As noted previously, in a volume compression incoherent granular soils undergo slight deformation. Unlike them, clay soils are characterized by a significant compressibility and lateral mobility. In other words, when soils are moist there is a sharp increase in the moisture coefficient of lateral pressure. Therefore, in addition to the grounds for such calculation on the strength and stability of the base, it is also necessary to make calculations on the HDT. Currently, such calculations are made using complex nonlinear deformation models of soil and numerical methods [17, 22].

3 Practical appendix of the theory of strength of soil when solving the applied problems of geotechnics

3.1. Comparison of the results of the laboratory model experiments and the results of the calculation

Example 1. Experiment with the horizontal position of the tray

It is required to determine the limiting value of the active resistance of the soil on the fence. The vertical and horizontal stress limit, obtained from the results of experiments, is $\sigma_1 = 200\ \kappa Pa$, $\sigma_3 = 18.5\ \kappa Pa$. Height of the side movable wall of the device H = 9 cm.

Given: soil density is $\gamma = 1.48\ \kappa H/m^3$; angle of internal friction is $\varphi = 39°$; $\theta = 39°$ and lateral pressure coefficient is $\xi = 0.1$. The lateral pressure σ_1 is determined according to the obtained experiments according to the equations (2.19, 2.28)

$$= 132 * \sin 39 = 83$$

$$P = \sqrt{\sigma_1^2 l^2 + \sigma_3^2 s^2} = 18.5\sqrt{0.5/0.01 + 0.5} = 132\ \text{кPa}$$

$$=$$

$$l = \theta^* = cos(39 + 6) = cos45 = 0.7,\ s = cos45 = 0.7$$

$$\sigma_1 = \sqrt{\left(\frac{P^2}{l^2}\right) - \sigma_3^2 \frac{s^2}{l^2}} = \sqrt{\frac{132^2}{0.5}} - 18.5^2 * 1 = 186\ \kappa Pa$$

$$\sigma_\theta = P\cos\theta = 132\cos 39 = 102.5\ \kappa Pa$$

$$\tau_\theta = P\sin\theta = 132\sin 39 = 83\ \kappa Pa$$

$$E_a^x = \sigma_3 H^2 = 18.5 * 0.09^2 = 0.15\ \text{кH}$$

According to the theory of Coulomb-Mohr in relation to the experiment

$$E_M^x = \frac{\gamma H}{2}\left(H^2 + 2H\frac{\sigma}{\gamma}\right)tg^2\left(45 - \frac{\varphi}{2}\right) =$$

$$= \frac{14.8 * 0.09}{2}\left(0.09^2 + 2 * 0.09\frac{200}{14.8}\right)0.22 = 0.36\ \kappa H$$

Calculated results are comparable with experimental data.

$$E_э^x = \sigma_3 * H^2 = 18.5 * 0.09^2 = 0.15\ \text{кН};\ E_a^x = 0.14\kappa H;\ E_M^x = 0.36\ \kappa H.$$

Example 2. Experiment with the horizontal position of the tray

It is required to determine the limiting passive load on the side wall of the tray (Figure 3.1). The vertical and horizontal stress limit, obtained from the results of experiments with weights on an inclined surface, is $\sigma_1 = 88.5\ \kappa Pa$ и $\sigma_3 = 9.2\ \kappa Pa$. Physical indicators of the soil are taken, as, for Example 1. According to the equation (2.17) we determine the value of the coefficient of lateral pressure of the soil corresponding to the limiting stress state, $\sigma_3 > \sigma_1$

$$\xi = tg\alpha_\xi = \frac{\sigma_1}{\sigma_3} = \frac{9.2}{88.5} = 0.1;\ \alpha_\xi = 7°$$

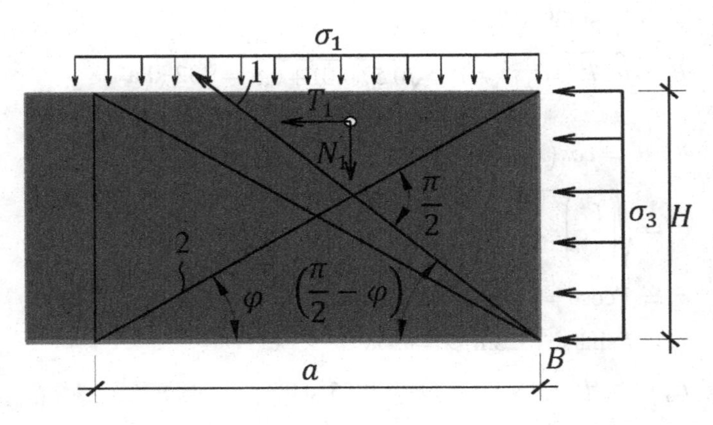

Figure 3.1 The settlement scheme of tests carried out with the active and passive loading. 1, 2 – shear surface.

According to equations (2.33) and (2.40) determine the value of the tangential and vertical limits of the stress required to keep the triangular prism in the condition of limit equilibrium:

$$\sigma_\theta = P\cos\theta = 75 * \cos 39 = 58 \text{ кPa}$$

$$\tau_\theta = P\sin\theta = 75\sin 39 = 467 \text{ кPa}$$

$$P = \sqrt{\sigma_1^2 l^2 + \sigma_3^2 s^2} = 9.2\sqrt{0.5/0.01 + 0.5} = 66 \text{ кPa}$$

$$=$$

$$l = \theta^* = cos(39 + 6) = cos45 = 0.7, s = cos45 = 0.7$$

$$\sigma_1 = \sqrt{\left(\frac{P^2}{l^2}\right) - \sigma_3^2\frac{s^2}{l^2}} = \sqrt{\frac{66^2}{0.5} - 9.2^2 * 1} = 93 \text{ кPa}$$

According to the theory of Coulomb-Mohr in relation to the experiment

$$\frac{\sigma_1}{\sigma_3} = tg^2\left(45 + \frac{\varphi}{2}\right); \ \frac{\sigma_1}{\sigma_3} = tg^2\left(45 + \frac{39}{2}\right) = 4.4$$

$$\sigma_1 = \sigma_3 tg^2\left(45 + \frac{\varphi}{2}\right) = 9.2 * 4.4 = 41 \ кPa$$

For comparison with the results of the experiment

$$E_э = \sigma_3 * H^2 = 41 * 0.09^2 = 0.32 \text{ кH and } \sigma_1 = 41 \ кPa$$

Example 3. An experiment with a tilted position of the tray (Figure 3.2)

It is proposed to do the calculation of the passive pressure of an inclined pad on a fixed wall. It is required to determine the strength of the soil layer inclined with respect to the horizon at an angle $\beta = 36°$, i.e. with b equal to the angle of repose of the ground. An inclined layer of soil is considered thick, $H = 9$ cm. The soil layer from the surface is not loaded. Compare the strength condition of the soil layer in a horizontal position and for the case of its oblique arrangement at an angle $\beta = 36°$.

Based on the global coordinate system, we define the main areas, with the direction of the perpendicular in relation to the slope:

$$\sigma_\theta = P\cos\theta = 8.2 * \cos 39 = 6.4 \text{ кPa}$$

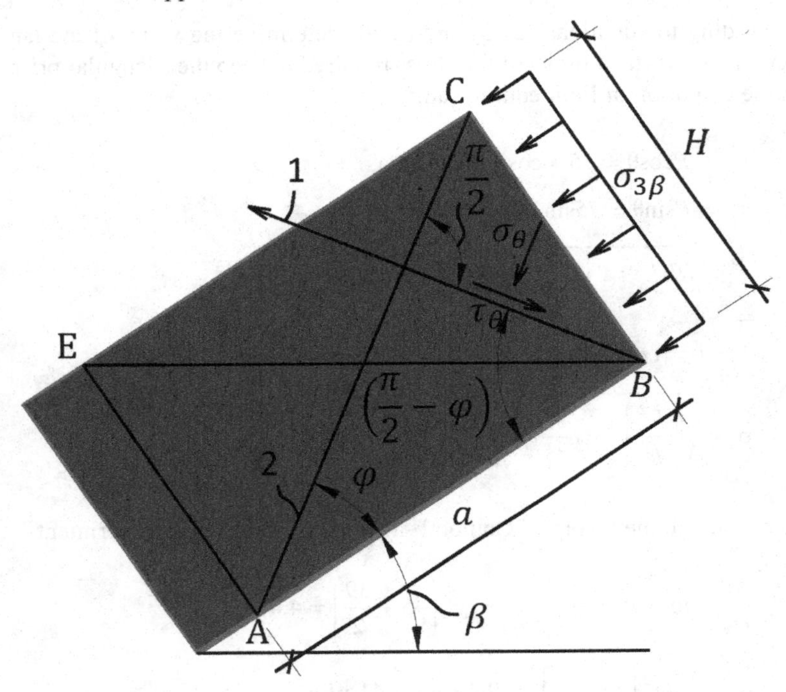

Figure 3.2 Driving tests. Diagram of inclined state of stress in the passive layer of soil loading.

1, 2 – shear surface.

$$\tau_\theta = P\sin\theta = 8.2\sin 39 = 5.1\ \text{кPa}$$

$$P = \sqrt{\sigma_1^2 l^2 + \sigma_3^2 s^2} = 1 * \sqrt{0.5 / 0.01 + 0.5} = 7.1\ \text{кPa}$$

$$l = \theta^* = cos(39+6) = cos45 = 0.7, s = cos45 = 0.7$$

$$\sigma_1 = \sqrt{\left(\frac{P^2}{l^2}\right) - \sigma_3^2 \frac{s^2}{l^2}} = \sqrt{\frac{7.1^2}{0.5} - 1^2 * 1} = 10\ \text{кPa}$$

$$\sigma_3 = \frac{\gamma H}{2cos\beta} = \frac{14.8 * 0.09}{2 * 0.8} \cong 1\ \kappa Pa$$

$$\text{E} = \sigma_1 * \text{H}^2 = 10 * 0.09^2 = 0.08\ \text{кН}.$$

3.2. Determination of active pressure on the retaining wall with various models of the theory of soil strength

Example 1

Calculation of active earth pressure on retaining wall with a vertical rear wall height H = 6 м. Given: soil-sand, density of compacted soil is γ = 18 kN/m³; angle of internal friction is $\varphi = 39°$.

Determine the magnitude of the vertical stress.

$$\sigma_1 = \gamma * H/2 + q = 18*6/2 + 0 = 54\ kPa$$

Let us compare the values of active pressure according to the Coulomb-Mohr theory of strength and according to the proposed option.

In accordance with the theory of Coulomb strength, the pressure on the vertical fence is the result of the shear forces of the elementary segments of the soil that are outside the slope (Figure 3.3). The calculation results of the geometric parameters of the slope are shown in Table 3.1.

The length l_i and area of S segments are defined by the equations

$$l_i \cong CH\left(\frac{\pi}{2n}\right)^k \qquad (3.1)$$

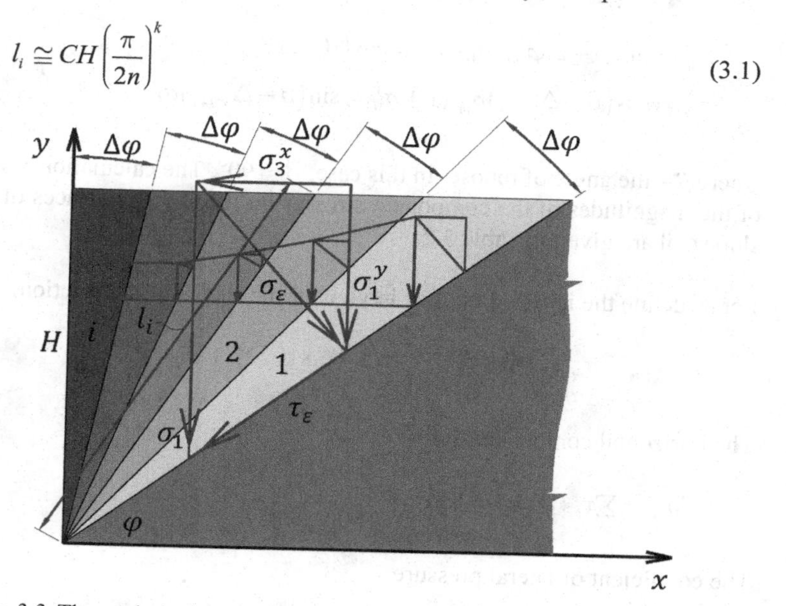

Figure 3.3 The analytical model of the slope to determine the coefficient of lateral pressure at rest, i.e. with the relative deformation ε = 0 and the critical load N on horizontal ground surface.

Table 3.1 Estimated values of the geometric parameters and weight of soil

The angle of the segment φ_i, (°)	*Height H, мм*	*k*	*Segment length* l_i, *мм*	*The area of the segment S, мм²*	*The specific weight of the soil γ, кН/м³*	*Weight of segment кН/м³*
39	6000	–0.82	10709.79	8 459 811	18	152
49	6000	–0.82	8881.636	6 006 927	18	108
59	6000	–0.82	7627.024	4 529 286	18	81
69	6000	–0.82	6708.069	3 561 413	18	64
79	6000	0.82	6003.431	3 500 846	18	63

where C = 35 and k = –0.82 equation parameters; n – the number of segments which divide the slope; and H – the height of the slope.

$$S = (l_{i-1} + l_i)^2 * tg(\varphi_i - \varphi_{i-1})/8 \tag{3.2}$$

Further, as shown in the scheme we define stress components for each segment:

$$\sigma_{\theta i} = \sin(\beta - \Delta\varphi_{i+1})\sigma_{1i};\ \tau_{\theta i} = cos(\beta - \Delta\varphi_{i+1})\sigma_{1i};$$

$$\sigma_{3i}^{x} = cos(\beta - \Delta\varphi_{i+1})\sigma_{\theta i} \text{ and } \sigma_{1i}^{y} = \sin(\beta - \Delta\varphi_{i+1})\sigma_{\theta i}$$

where β – the angle of repose. In this case, $\beta = 90°$. The calculation results of the magnitudes of the component stresses on the sliding surfaces of the slope soil are given in Table 3.2.

Let us define the inclined voltage in view of the shear surface friction:

$$\tau_i = \tau_{\theta i} - \sigma_{\theta i} * tg\varphi = 46.2 - 28.3 * 0.81 = 23.3\ kPa$$

The horizontal component of it

$$\Sigma\sigma_{xi} = \Sigma\tau_i * cos\varphi i = 8\ kPa$$

The coefficient of lateral pressure

$$\xi = \frac{\sigma_{xi}}{\sigma_1} = \frac{8.0}{56.1} = 0.14$$

Table 3.2 Estimated values of stress components on the surfaces of the sliding soil

Segment number	*The angle of the segment (°)*		*The values of stress components for a particular segment (kPa)*				
	$\Delta\varphi_i$	β	σ_1	σ_θ	τ_θ	σ_3^x	σ_1^y
1	39	90	12.6	9.8	7.9	6.2	7.6
2	48	90	7.6	5.1	5.6	3.8	3.4
3	54	90	8.0	4.7	6.5	3.8	2.8
4	61	90	7.3	3.5	6.4	3.1	1.7
5	68	90	6.7	2.5	6.3	2.3	0.9
6	75	90	6.3	1.6	6.1	1.6	0.4
7	82	90	7.6	1.1	7.5	1.0	0.1
	90	90	56.1	28.3	46.2	21.8	17.0

Active earth pressure on the fence

$$E_a^x = N\ \sigma_{xi} = 6 * 8 = 48\ kN$$

For comparison, look at the active earth pressure on the fence in accordance with the theory of strength of Mohr-Coulomb:

$$E_a = \frac{\gamma N^2}{2} tg^2\left(45 - \frac{\varphi}{2}\right) = \frac{18 * 6^2}{2} tg^2\left(45 - \frac{39}{2}\right) = 74\ kN$$

Below we consider the definitions of these stresses by the proposed method of (2.27, 2.28). $\alpha_\xi = 7°$ and $\xi = 0.12$. Let us determine the magnitude of the ultimate normal stresses at shear pads deviated from the horizontal by an angle

$$\varphi^* = 39 + 7 = 46°$$

$$\sigma_\theta = P\cos\theta = 38.4 * \cos 39 = 29 \text{ кPa}$$

$$\tau_\theta = P\sin\theta = 38.4 * \sin 39 = 24.2 \text{ кPa}$$

кПа

$$P = \sqrt{\sigma_1^2 l^2 + \sigma_3^2 s^2} = 54 * \sqrt{0.5 + 0.5 * 0.01 +} = 38.4 \text{ кPa}$$

=

$$l = \theta^* = cos(39 + 6) = cos35 = 0.82, s = cos55 = 0.57$$

$$\sigma_3 = \sqrt{\left(\frac{P^2}{s^2}\right) - \sigma_1^2 \frac{l^2}{s^2}} = \sqrt{\frac{38.4^2}{0.5} - 54^2 * 1} = 5.7 \text{ кPa}$$

$$\sigma_1 = \gamma * H / 2 = 18 * 6 / 2 = 54 \, \kappa Pa$$

Checking the strength of the soil condition on the surface of the shear $\frac{\tau_\theta}{\sigma_\theta} = \frac{34}{42} = 0.81 = tg\varphi$, or $\varphi = 39°$. We define an active earth pressure on retaining wall

$$E_a = N \, \sigma_3^x = 6 * 7 = 42 \, kN.$$

Example 2

Calculation example given in the book [3]. Determine the active pressure on the shear surfaces BC, including the retaining wall with a height of 5 m. The angle of internal friction of soil $\varphi = 35°$, the coefficient of lateral pressure $\xi = 0.20$, the unit weight of soil $\gamma = 15.3$ kN/m^3. The calculation results are stress components at the shear surfaces and by the method of the authors, presented in Table 3.3 and Table 3.4.

$$\sigma_1 = \frac{\gamma N}{2} = 5 * \frac{15.3}{2} = 38 \; kPa.$$

We identify the inclined voltage in view of the shear surface friction:

$$\tau_i = \tau_{\theta i} - \sigma_{\theta i} * tg\varphi = 35.8 - 24 * 0.7 = 19 \; kPa$$

Table 3.3 Calculating the geometry and the weight of soil

The angle of the segment φ_i (°)	*Height H, мм*	k	*The length of segment l_i, мм*	*The area of segment S, мм^2*	*The specific weight of the soil γ, кН/м^3*	*Weight of segment кН*
35	5000	–0.82	9482	5 374 754	15.3	82.2
43	5000	–0.82	8009	3 938 407	15.3	60.3
51	5000	–0.82	6964	3 034 530	15.3	46.4
59	5000	–0.82	6179	2 562 943	15.3	39.2
67.5	5000	–0.82	5534	2 081 015	15.3	31.8
76	5000	–0.82	5021	1 634 570	15.3	25.0
84	5000	–0.82	4625	1 483 211	15.3	22.7
Total						307

Table 3.4 The calculated values of stress components on the surfaces of the sliding soil

Segment number	*The angle of the segment (°)*		*The values of stress components for a particular segment (kPa)*				
	$\Delta\varphi_i$	β	σ_1	σ_θ	τ_θ	σ_3^x	σ_1^y
1	35	90	8.7	7.1	5.0	4.1	5.8
2	43	90	7.5	5.5	5.1	3.8	4.0
3	51	90	6.7	4.2	5.2	3.3	2.6
4	59	90	6.3	3.3	5.4	2.8	1.7
5	67.5	90	5.8	2.2	5.3	2.0	0.8
6	76	90	5.0	1.2	4.8	1.2	0.3
7	84	90	4.9	0.5	4.9	0.5	0.1
	90	90	44.9	24.0	35.8	17.6	15.4

and its horizontal component

$$\sum\sigma_{xi} = \sum\tau_i * cos\varphi i = 6.8\ kPa$$

The coefficient of lateral pressure

$$\xi = \frac{\sigma_{xi}}{\sigma_1} = \frac{6.8}{38} = 0.18$$

Active earth pressure on a vertical rail is calculated according to this method:

$$E_a^x = N\sigma_3^x = 6.8 * 5 = 34\,kH\ N$$

The solution to this problem by the method proposed by the authors:

$$\sigma_\theta = P\cos\theta = 38.4 * \cos 39 = 29\ \text{кPa}$$

$$\tau_\theta = P\sin\theta = 38.4 * \sin 39 = 24.2\ \text{кPa}$$

кPa

$$P = \sqrt{\sigma_1^2 l^2 + \sigma_3^2 s^2} = 38 * \sqrt{0.47 + 0.53 * 0.04} = 27\ \text{кPa}$$

$=$

$$l = \theta^* = cos(35 + 11.5) = cos46.5 = 0.688,\ s = cos43.5 = 0.72$$

$$\sigma_3 = \sqrt{\left(\frac{P^2}{s^2}\right) - \sigma_1^2 \frac{l^2}{s^2}} = \sqrt{\frac{27^2}{0.53} - 38^2 * \frac{0.47}{0.53}} = 9.7 \text{ кPa}$$

$$E_a = H\sigma_3 = 5 * 9.7 = 48 \ \kappa H$$

Active earth pressure on the fence in accordance with the theory of strength of Mohr-Coulomb is:

$$E_a^x = \frac{\gamma H^2}{2} tg^2\left(45 - \frac{\varphi}{2}\right) = \frac{15.3 * 5^2}{2} tg^2\left(45 - \frac{35}{2}\right) = 52 \ kN$$

3.3. Determination of active pressure on the retaining wall with complex geometric configuration of the hillside

When solving the following Task, the influence of the calculation scheme on the final calculation result is determined. As is known, to solve the problem of determining the ground pressure on the fence, traditionally we take as the basis the normal stresses acting on the horizontal platform that runs at the line of the foundation. To solve some problems in the beginning it is more convenient to calculate the voltage limiting σ_φ and τ_φ acting on the court, off the horizontal at an angle $\theta = \varphi^*$, and then determine other design characteristics. Below are some examples of determining the soil pressure on the wall of protective skin, taking into account the loaded slope with an intense load q. This would be considered a more general case, when the rear wall of the retaining wall has a slope of δ.

Initial data for the calculation: the back wall of the retaining wall is inclined from the vertical by an angle δ, and the angle of deviation from the soil surface horizon is β; the angle of internal friction of soil is φ; the lateral pressure coefficient is ξ; the height of the wall is equal to N. We assume that the soil behind the retaining wall is homogeneous, with a density equal to the addition of γ. It is required to determine the voltage on the retaining wall.

First, define the following geometric dimensions: length of the rear wall of the retaining wall $L_1 = N/\sin\delta$; and the depth of the oblique line perpendicular to the sliding surface sliding wedge (Figure 3.4):

$$N^2 = \sin\left(\pi - \varphi^* - \delta\right) L^1 = H * \sin\left(\pi - \varphi^* - \delta\right) / \sin\delta \qquad (3.3)$$

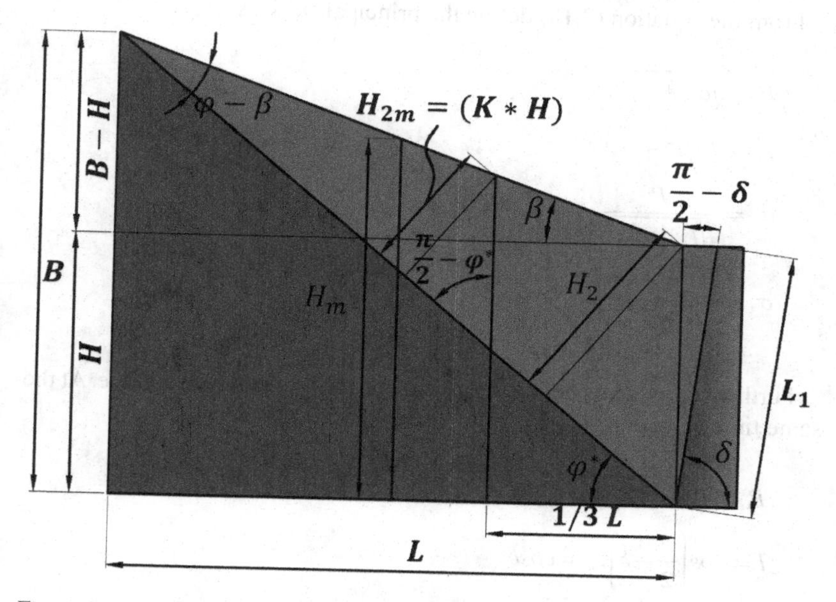

Figure 3.4 The analytical model of the retaining wall and slope.

The average depth of the oblique line perpendicular to the sliding surface sliding wedge

$$N_m = \frac{N_2}{2} = \frac{N}{2} * \frac{sin\left(\pi - \varphi^* - \delta\right)}{sin\delta} \tag{3.4}$$

We determine the stress components σ_θ and τ_θ on the shear surface of the collapse prism $\theta = \varphi^* = \varphi + \alpha_\xi$

$$\sigma_\theta = \gamma N_m + qcos\varphi^* = \frac{\gamma N}{2} * \frac{sin\left(\pi - \varphi^* - \delta\right)}{sin\delta} + qcos\varphi^*$$

and

$$\tau_\theta = tg\varphi\sigma_\theta = tg\varphi\left[\frac{\gamma N}{2} * \frac{sin\left(\pi - \varphi^* - \delta\right)}{sin\delta} + qcos\varphi^*\right] \tag{3.5}$$

From the equation (2.18) define the principal stresses:

$$P=\sqrt{\sigma_{\theta}^{2}+\tau_{\theta}^{2}}$$

$$P=\sqrt{\sigma_{1}^{2}l^{2}+\sigma_{3}^{2}s^{2}}$$

$$\sigma_{1}=\frac{P}{\sqrt{l^{2}+\xi^{2}s^{2}}}$$

$$\sigma_{3}=\frac{P}{\sqrt{l^{2/}\xi^{2}+s^{2}}}$$

Further components define the voltage on the inclined wall surface. At the same time we accept $\theta=\delta-\alpha_{\xi}$

$$P_{\delta}=\sqrt{\sigma_{1}^{2}l^{2}+\sigma_{3}^{2}s^{2}}$$

$$l=cos\left(\frac{\pi}{2}-\delta\right), s=cos\delta$$

$$\sigma_{\theta}=\sigma_{1}l^{2}+\sigma_{3}s^{2}$$

$$\tau_{\theta}=\sqrt{P_{\delta}^{2}-\sigma_{1}^{2}}$$

The active force on the retaining wall is defined by the equation

$$E_{a}=\sigma_{\theta}*L_{1}$$

Now we are going to solve the same problem for another settlement scheme. To do this, we need to determine the average depth of the prism of oblique collapse homogeneous formation. We define the length of the base of the prism of collapse.

$$l=\frac{N}{tg\varphi^{*}-tg\beta} \tag{3.6}$$

The average depth of the prism of collapse to the level line of the foundation of the retaining wall is:

$$N_{m}=l\frac{tg\beta}{2}+N=\frac{N}{tg\varphi^{*}-tg\beta}\frac{tg\beta}{2}+N=N\left[\frac{tg\beta}{2\left(tg\varphi^{*}-tg\beta\right)}+1\right] \tag{3.7}$$

Next, we define the main voltage corresponding to the global coordinate system.

$$\sigma_1 = \frac{\gamma H_m}{2}$$

$$P_\delta = \sigma_1\sqrt{l^2 + \xi^2 s^2}$$

$$\sigma_3 = \frac{P}{\sqrt{l^{2/}\xi^2 + s^2}}$$

For area $\theta = (\delta)$

$$l = cos\delta, s = cos\left(\frac{\pi}{2} - \delta\right),$$

$$\sigma_\theta = \sigma_1 l^2 + \sigma_3 s^2$$

$$P_\delta = \sqrt{\sigma_1^2 l^2 + \sigma_3^2 s^2}$$

$$\tau_\theta = \sqrt{P^2 - \sigma_\theta^2}$$

The active force acting on the retaining wall is defined by the equation

$$E_a = \sigma_{3\theta} * L_1$$

Example 3. Calculation of active earth pressure on retaining wall

Initial data for the calculation: H = 6 m, $\delta = 82° \varphi = 28°$, $\beta = 20°$, $\alpha_\xi = 12°$, $\xi = 0.21$, and $\gamma = 18\frac{kH}{M^3}$, $q = 0$. The geometric scheme of the retaining walls and slope is shown in Figure 3.4.

In accordance with the equation (2.24) we define the angle of deflection of the shear area, and the principal stresses acting on this site:

$$\theta = \varphi^* = \varphi + \alpha_\xi = 28 + 12 = 40°$$

Weight sliding wedge G

$$G = \frac{\gamma H_2^2}{2}\left[\frac{1}{tg\left(\varphi^* - \beta\right)} + \frac{1}{tg\left[\pi - \left(\varphi^* + \delta\right)\right]}\right] =$$

$$\frac{18*5.1^2}{2}\left[\frac{1}{tg(40-20)} + + \frac{1}{tg\left[\pi - (40+82)\right]}\right] = 778kN$$

$$\sigma_\theta = \frac{G}{l} cos\varphi^* = Gcos\varphi^* \frac{\left(tg\varphi^* - tg\beta\right)}{H} = 778*0,766\frac{0.475}{6} = 47.2\,\text{kPa}$$

$$\text{Or } N_2 = H * \frac{sin\left(\pi - \varphi^* - \delta\right)}{sin\delta} = 6\frac{sin\left(\pi - 40 - 82\right)}{sin82} = 5.2\,\text{м}$$

$$\sigma_\theta = \frac{\gamma N_2}{2} = \frac{18*5.2}{2} = 47\text{ kPa}$$

Normal and tangential stresses acting on the playground, the deviation from the horizontal line at an angle $\theta = \varphi^*$

$$P = \sqrt{\sigma_\theta^2 + \tau_\theta^2} = \sqrt{62.4^2 + 33.2^2} = 53\ \kappa Pa$$

$$l = cos\left(28 + 12\right) = 0.766,\ s = cos50 = 0.64$$

$$\sigma_1 = \frac{P}{\sqrt{l^2 + \xi^2 s^2}} = \frac{53}{\sqrt{0.59 + 0.044*0.41}} = 60.4\ \text{кPa}$$

$$\sigma_3 = \frac{P}{\sqrt{l^{2/}\xi^2 + s^2}} = \frac{53}{\sqrt{0.59/0.044 + 0.41}} = 14.3\ \text{кPa}$$

Orientation wall of surface area is equal to:

$$\delta = 82°$$

whence the magnitude of normal and tangential stresses on the rear inclined platform of the retaining wall

$$\sigma_\theta = \sigma_1 l^2 + \sigma_3 s^2 = 60.4*0.02 + 14.3*0.98 = 15\ \text{кPa}$$

$$P = \frac{\sigma_\theta}{\cos\theta} = \frac{15}{\cos 28} = 17\ \text{кPa}$$

$$P_\delta = \sqrt{\sigma_1^2 l^2 + \sigma_3^2 s^2} = \sqrt{3648*0.02 + 204*0.98} = 17\ \text{кPa}$$

$$\tau_\theta = \sqrt{P^2 - \sigma_\theta^2} = \sqrt{273 - 213} = 8\ \kappa Pa$$

Active force acting on retaining the wall

$$E_a = \sigma_\theta * L_1 = 15*\frac{6}{0.99} = 91\ \text{кН}.$$

Now we will solve the same problem for another calculation scheme traditionally used in the literature [19, 17]. To do this, we determine the average depth of the collapse prism of an inclined homogeneous reservoir. Base length of the collapse prism:

$$l = \frac{H}{tg\varphi^* - tg\beta} \tag{3.8}$$

The average depth of the collapse prism to the level of the bottom of the foundation of the retaining wall

$$H_m = l\frac{tg\beta}{2} + H = \frac{H}{tg\varphi^* - tg\beta}\frac{tg\beta}{2} + H = H\left[\frac{tg\beta}{2\left(tg\varphi^* - tg\beta\right)} + 1\right] \tag{3.9}$$

$$H_m = H\left[\frac{tg\beta}{2\left(tg\varphi^* - tg\beta\right)} + 1\right] = 6\left[\frac{tg20}{2(tg40 - tg20)} + 1\right] = 8.2 \text{ м.}$$

We define the main stresses corresponding to the global coordinate system $\theta = \varphi^*$

$$\sigma_1 = \frac{\gamma H_m}{2} = \frac{18 * 8.2}{2} = 73.8 \; \kappa Pa$$

$$P_\delta = \sigma_1\sqrt{l^2 + \xi^2 s^2} = 73.8\sqrt{0.59 + 0.044 * 0.41} = 57 \text{ кPa}$$

$$\sigma_3 = \frac{P}{\sqrt{l^{2/}\xi^2 + s^2}} = \frac{57}{\sqrt{0.59/0.044 + 0.41}} = 15.3 \text{ кPa}$$

Compound stresses at the shear site

$$\sigma_\theta = P\cos\theta = 57\cos 28 = 50.3 \text{ кPa}$$

$$\tau_\theta = P\sin\theta = 57\sin 28 = 26.7 \text{ кPa}$$

The orientation of the surface area of the wall is

$$\delta_\xi = 82°$$

which gives the next

$$\sigma_\theta = \sigma_1 l^2 + \sigma_3 s^2 = 73.8 * 0.02 + 15.3 * 0.98 = 16.5 \text{ кPa}$$

Active force acting on the retaining wall

$$E_a = \sigma_{3\theta} * L_1 = 16.5 * \frac{6}{0.99} = 100 \text{ кН}.$$

As can be seen from the presented examples, the discrepancy in the final results is 100/91 = 1.21 times. This difference is associated with the averaged determination of the vertical principal stresses. σ_1 according to the second calculation scheme.

For comparison, the active pressure of the soil is determined by the Rankine formula [3]. $K_{a(R)} = 0.55$

$$E_{a(R)} = \frac{1}{2} H^2 \gamma K_{a(R)} = 0.5 * 36 * 18 * 0.55 = 178 \text{ кPa}$$

Example 4

Calculation of the active soil pressure on the retaining wall from the slope (cover soils on a rocky base) loaded with an intensive load q = 5 kN/m. The retaining wall height H is 8.4 m. (Figure 3.5). The calculated characteristics of the soil backfill: $\gamma = 18$ kN/ m^3; $\varphi = 28°$; $\delta = 0$; $\beta = 25°$; $\xi = 0.2$ and $\alpha_\xi = 11°$.

At the beginning of the calculation, we determine the deviation angle of the most dangerous pad in relation to the horizon $\varphi^* = 28° + 11° = 39°$. Note that on this platform there is the normal and tangential $\tau_\varphi \sigma_\varphi$ voltage. Maximum thickness of the sliding wedge:

$$b = H \sin\left(\frac{\pi}{2} - -\varphi\right) = 8.4 * \sin(90 - 38) \cong 6.5 \text{ м}.$$

Note that the estimated shear surface exceeds the angle, i.e. $\theta^* = 39° > \beta = 25°$

$$\sigma_1 \cong \frac{\gamma H}{2} + q = \frac{18 * 8.4}{2} + 5 = 80.6\, кPa$$

Determine the normal shear stresses on the shear surface:

$$l = \cos(39) = 0.777, s = cos51 = 0.63$$

$$\sigma_\theta = \sigma_1 l^2 + \sigma_3 s^2 = 80.6(0.6 + 0.04 * 0.4) = 50 \text{ кPa}$$

$$\tau_\varphi = \sigma_\varphi tg\varphi = 50 * 0.53 = 33.2\, кPa$$

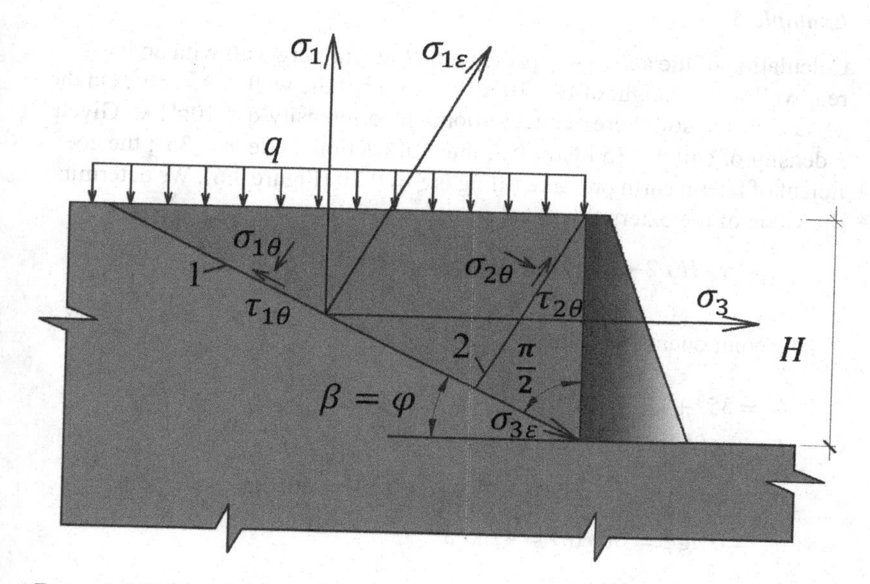

Figure 3.5 Design diagram of a retaining wall perceiving active pressure from soil and distributed load q

by the equations (2.18) and (2.27) define σ_1, σ_3. Calculations are made for the angle $\theta = \varphi = 38°$.

according to equations (2.18) and (2.27) define σ_1, σ_3. Make up the calculating of angles $\theta^* = 39°$ and

$$P = \frac{\sigma_\theta}{\cos\theta} = \frac{50}{\cos 28} = 56\ \kappa Pa$$

$$\sigma_1 = \frac{P}{\sqrt{l^2 + \xi^2 s^2}} = \frac{56}{\sqrt{0.6 + 0.04 * 0.4}} = 71.3\ \kappa Pa$$

$$\sigma_3 = \frac{P}{\sqrt{l^{2/}\xi^2 + s^2}} = \frac{56}{\sqrt{0.6 / 0.04 + 0.4}} = 14.3\ \kappa Pa$$

$$E_{3\beta} = \sigma_3 * h = 14.3 * 8.4 = 120\ \kappa H.$$

For $\beta = 25°$ no shift on the contact surface

Example 5

Calculation of the active soil pressure on the retaining wall with an inclined rear wall with a height of H = 10m. The slope of the wall is $\delta = 80°$. On the surface of the soil there is an additional load intensity q = 10кН/м. Given: a density of soil γ = 15 kN/m ^ 3; internal friction angle $\varphi = 35°$; the coefficient of lateral earth pressure filling is $\xi = 0.20$ (Figure 3.6). We determine the value of the external vertical stresses.

$$\sigma_1 = \gamma * H / 2 + q = 15 * 10 / 2 + 10 = 85 \text{ kPa}$$

The components of voltage

$$\theta^* = 35° + 11° = 46°$$

$$l = \cos(46) = 0.69, s = cos44 = 0.72$$

$$\sigma_\theta = \sigma_1 l^2 + \sigma_3 s^2 = 85(0.47 + 0.04 * 0.53) = 60 \text{ кPa}$$

$$\tau_\varphi = \sigma_\varphi tg\varphi = 60 * 0.7 = 42 \text{ к}Pa$$

$$P = \frac{\sigma_\theta}{\cos\theta} = \frac{60}{\cos 35} = 73 \text{ к}Pa$$

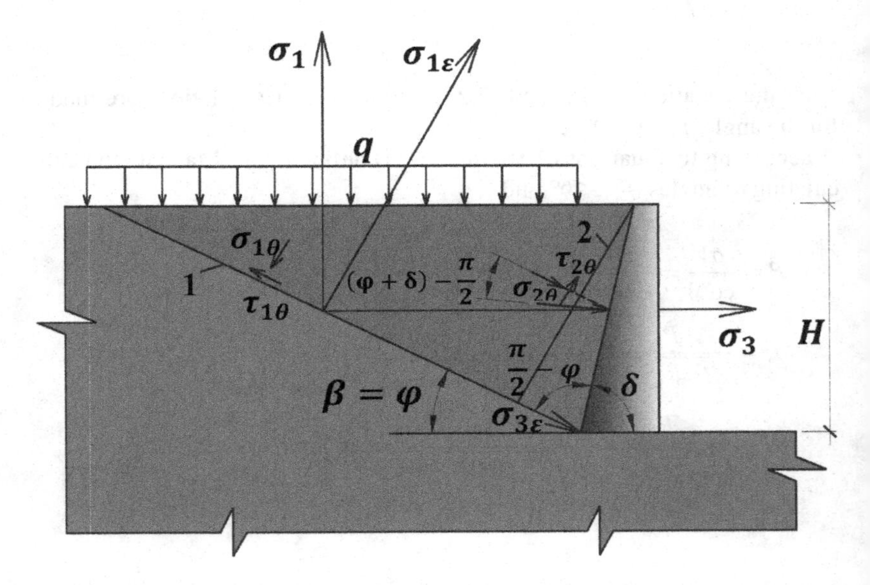

Figure 3.6 Design diagram of a retaining wall with an inclined rear wall perceiving active pressure from the ground and distributed load q.

$$\sigma_1 = \frac{P}{\sqrt{l^2 + \xi^2 s^2}} = \frac{73}{\sqrt{0.47 + 0.04 * 0.53}} = 104 \ \kappa Pa$$

$$\sigma_3 = \frac{P}{\sqrt{l^{2/}\xi^2 + s^2}} = \frac{73}{\sqrt{0.47 / 0.04 + 0.53}} = 21 \kappa Pa$$

$$\sigma_\theta = \sigma_1 l^2 + \sigma_3 s^2 = 21(0.47 + 0.04 * 0.53) = 60 \ \kappa Pa$$

$E_a^x = H\sigma_\theta = 10 * 31.1 = 311 kN$ force direction perpendicular to the surface of the wall.

Example 6

We determine confining pressure on the mound height H = 12 m. Acting on the structure of the underground tunnel the measurement section is b = 4 m, h_o = 6m. The main design characteristics of soil bulk density of soil γ is 19 kN/m³; internal friction angle $\varphi = 35°$. The value of the coefficient of lateral earth pressure filling the same is:

$\xi = 0.20.$

Assume that the tunnel is located within a stable triangular mound. The angle of deflection limiting shear surface with respect to the horizon is assumed to be $\theta^* = \varphi + \alpha_\xi$. Figure 3.7 shows the design scheme indicating the shear surface.
Determine the height of the arch collapses

$$h_1 = \frac{tg\varphi^* b}{2} = \frac{tg(35 + 11) * 4}{2} = 2.1 \ \text{м}.$$

and the $h_2 = \text{H} - (h_o + h_1) = 12 - (6 + 2.1) = 3.9$ м.m.

Define a confining pressure at the points d, b and c^1

$$\sigma_{gd} = \gamma h_2 = 19 * 3.9 = 74.1 \ kPa.$$

$$\sigma_{gb} = \gamma \ (h_1 + h_2) = 19 * (2.1 + 3.9) = 114 \ kPa.$$

$$\sigma_{gc^1} = \gamma H = 19 * 12 = 228 \ kPa.$$

The values of the average gravitational pressure acting on the inclined surface of the slope:

$$\sigma_{gd} = \frac{74.1}{2} = 37 \ kPa$$

$$\sigma_{gb} = \frac{74.1+114}{2} = 94\ kPa$$

$$\sigma_{gc^1} = \frac{228+114}{2} = 171 kP$$

We define the projection of gravitational forces on the normal and tangential values to the inclined surface of the slope:

$$\sigma^n_{gb} = \sigma_{gb} cos\varphi^* = 94 * 0.69 = 65\ kP$$

$$\sigma^n_{gc^1} = \sigma_{gc^1} cos\varphi^* = 171 * 0.69 = 118\ kP$$

$$\tau_{gb} = \sigma_{gb} sin\varphi^* = 94 * 0.72 = 685\ kP$$

$$\tau_{gc^1} = \sigma_{gc^1} sin\varphi^* = 171 * 0.72 = 123\ kP$$

Determine the magnitude of the total voltage vector P

$$P = \sqrt{\sigma_\theta^2 + \tau_\theta^2} = \sqrt{118^2 + 123^2} = 170 \text{ кПа}$$

$$\sigma_3 = \frac{P}{\sqrt{l^{2/}\xi^2 + s^2}} = \frac{170}{\sqrt{0.48/0.04 + 0.52}} = 48 \text{ кПа}$$

$$l = cos(35+11) = cos46 = 0.69 \text{ и } s = cos44 = 0.72$$

Determine the active load of the soil on the side wall of the tunnel from the lateral spread of a triangular prism (I).

From the equation (2.18) we define an active earth pressure on the side wall of the tunnel:

$$\sigma_3 = \sigma_\theta \frac{sin\alpha_\xi}{cos\theta} = 118 \frac{0.19}{0.82} = 27.3\ kPa$$

We define a resistive load of soil on the side wall of the tunnel from the side thrust of triangular prism (I):

$$T_I = \frac{\gamma h_0^2}{tg\varphi^*} \xi = \frac{19*36}{1} 0.2 = 136 \text{ кН}$$

The total resistive load on the horizontal wall of the tunnel

$$\Sigma T_I = 27.3 * 6 + 136 = 289 \ \kappa H$$

Vertical earth pressure tunnel to cover the prism of the collapse of the IV is determined by the equation (2.27).

Given that:

$$\sigma_\theta = \sigma''_{gb}$$

$$P = \frac{\sigma_\theta}{\cos\theta} = \frac{65}{0.7} = 93 \ \kappa\Pi a$$

We define a mountain of earth pressure on the tunnel to cover of the triangular arch abd (II):

$$N_{II} = 2\gamma\frac{b * h_1}{4} = 2\gamma\frac{b^2 * tg\varphi^*}{8} = 2 * 19\frac{16 * 1.03}{8} = 78.3 \text{ kN}$$

$$\sigma_{II} = \frac{N_{II}}{b} = \frac{78.3}{4} = 19.5 \text{ kPa}$$

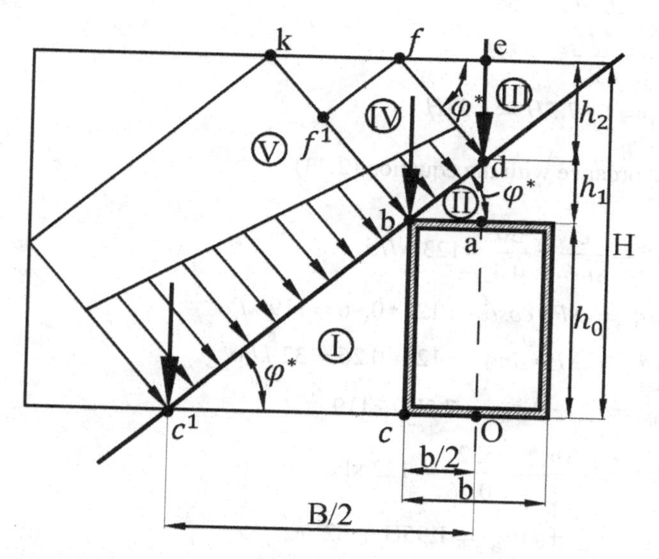

Figure 3.7 Design scheme for the determination of rock pressure on the vault (cover) and a vertical tunnel wall, located inside the ground cloth (mound).

The total pressure on the horizontal surface of the coating tunnel

$$\sum \sigma_1 = \sigma_1 + \sigma_{II} = 91 + 19.5 = 110.5 \text{ kPa}$$

In the direct determination of earth pressure on equation

$$\sum \sigma_1 = \gamma(h_1 + h_2) = 19(2.1 + 3.9) = 114 \text{ kPa}$$

It follows that during the laying of the tunnel in the mound (ground cloth), the pressure of soil on the surface of the coating can be defined as the product of the proportion of the soil to the depth.

3.4. Solving problems and determining the slope stability of the active pressure on the retaining wall with respect to cohesive soils

Example 7

Calculating the marginal height of the slope. Given: soils of little wet loam soil density $\gamma = 15$ kN/m^3; internal friction angle $\varphi = 24°$; specific cohesion $c = 50$ kN/m^2; lateral pressure coefficient

$$\xi = 0.26; \alpha\xi = 15°.$$

$$\sigma_{1f} = \frac{\gamma H}{2} = 7.5H$$

$$\sigma_{3f} = \xi * 7.5H = 1.95H$$

In accordance with the equation (2.33)

$$\Delta P = \frac{c}{sin\theta} = \frac{50}{0.4} = 123 \ \kappa\Pi a$$

$$\Delta\sigma_{1c} = \Delta P * cos\alpha_{\xi} = 123 * 0.96 = 119 \ kPa$$

$$\Delta\sigma_{3c} = \Delta P * sin\alpha_{\xi} = 123 * 0.26 = 32 \ kPa$$

$$\sigma_{1c} = \sigma_{1\varphi} + \Delta\sigma_{1c} = 7.5\text{H} + 119$$

$$\Delta\sigma_{1c} = \frac{\Delta\sigma_{3c}}{\xi} = \frac{32}{0.26} = 122 \text{ кPa}$$

$$\sigma_{3c} = \sigma_{3\varphi} + \Delta\sigma_{3c} = 1.95\text{H} + 32$$

So as $\sigma_{3c} = \xi\sigma_{1c} \cong c = 50 \ \kappa Pa$

$$\sigma_{1c} = \frac{c}{\xi} = \frac{50}{0.26} = 192 \text{ кПа}$$

$$\sigma_{1c} = \sigma_{1\varphi} + \Delta\sigma_{1c} = 7.5\text{H} + 119 = 192 \text{ кPa}$$

$$\text{H} = 9.7 \text{ м.}$$

$$\sigma_{3c} = \sigma_{3\varphi} + \Delta\sigma_{1c} = 1.95\text{H} + 32 = 50 \text{ кPa}$$

Thus, the maximum height of the slope H, taking into account the proposed theory, can be determined by the equation

$$H = \frac{2c}{\gamma}\left(\frac{1}{\xi} - \frac{cos\alpha_{\xi}}{sin\varphi}\right) \tag{3.10}$$

For comparison, the maximum height of the vertical slope, calculated by the equation [17]

$$H = \frac{2c}{\gamma}\frac{cos\varphi}{(1 - sin\varphi)} \tag{3.11}$$

$$H = \frac{2 * 50 * 0.91}{15(1 - 0.4)} = 10.2 \text{ м.}$$

Example 8

Calculation of slope stability, limit the surface loaded with a distributed load of intensity q. The height of the slope H = 10 m. Loam soil with soil density $\gamma = 17$ kN/m^3; internal friction angle $\varphi = 25°$; specific cohesion c = 15 kPa; lateral pressure coefficient $\xi = 0.35$. The angle of repose of hard prism $\theta = 30°$. It is required to determine the critical distributed load q.

The average vertical stress of the right side of the thrust prism is:

$$\sigma_{1f} = \frac{\gamma H}{2} + q = 17 * \frac{10}{2} + q = 85 + q$$

For the case of active loading of structured soils, by the equation (2.48) we determine the stress components: $\varphi = 25°$

$$\sigma_{1c} = \sigma_{1\varphi} + \frac{c * cos\alpha_{\xi}}{sin\varphi} = 136 + q$$

$$\sigma_{3c} = \sigma_{3\varphi} + \Delta\sigma_{3c} = 0.35(85+q)+11.5 = 41.2+0.35q$$

$$P = \sigma_{1c}\sqrt{l^2+\xi^2 s^2} = (136+q)\sqrt{0.44+0.12*0.56} = 0.71\ (136+q)$$

$$l = \cos(19+30) = 0.66, s = \cos\left(\frac{\pi}{2}-\theta^*\right) = 0.75$$

$$\sigma_\theta = P\cos\theta = 0.71(136+q)*0.91 = 0.64(136+q)$$

$$\tau_\theta = P\sin\theta = 0.3(136+q)$$

Let us determine the stability of the bearing prism of a shear load. The stability of the supporting triangular prism, which perceives lateral pressure, depends on the condition of the limiting resistance of the soil at its base. Normal stresses at the base of the slope:

$$\sigma_{1f} = \frac{\gamma H}{2} = \frac{17*10}{2} = 85\ \kappa Pa$$

$$\tau_\theta = \sigma_1 tg\varphi + c = 85*0.466+15 = 54.61\ \kappa Pa$$

For the right prism collapse $\sigma_{3c} = \tau_\theta = 54.61\ \kappa Pa$. Make an equation of equilibrium

$$\sigma_{3c} = 41.2+0.35q = 54.61\ \kappa Pa$$

$$q = 39\ \kappa Pa$$

Example 9

Calculation of active earth pressure on retaining wall. Initial data for calculation: h = 8 m, $\delta = 90°$, $\varphi = 22°$, $\beta = 0°$,

$\alpha_\xi = 19°$, $\xi = 0.35$ and $\gamma = 18$ kH/m^3, and c = 12 kPa.

Solution 1

We determine the average pressure of the vertical column of soil located behind the retaining wall:

$$\sigma_{1f} = \frac{\gamma H}{2} = \frac{18*8}{2} = 72\ kPa$$

$$\sigma_{1c} = \sigma_{1\varphi} + \frac{c * cos\alpha_{\xi}}{sin\varphi} = 72 + \frac{12 * 0.95}{0.37} = 103 \text{ кPa}$$

We define active pressure-structured soils on the fence:

$$P = \sigma_{1c}\sqrt{l^2 + \xi^2 s^2} = 102\sqrt{0.56 + 0.12 * 0.44} = 80$$

From whence

$$l = \cos(19 + 22) = 0.75, s = \cos\left(\frac{\pi}{2} - \theta^*\right) = 0.66$$

Determine the pressure of the soil on the fence:

$$\sigma_{3c} = \sigma_{3\varphi} + \Delta\sigma_{3c} = \xi\sigma_{1f} + \text{c}\frac{sin\alpha_{\xi}}{sin\theta} = 0,3 * 72 + 12\frac{0.29}{0.37} = 31 \text{ кPa}$$

Define the maximum height of a stable slope:

$$H = \frac{2c}{\gamma}\left(\frac{1}{\xi} - \frac{cos\alpha_{\xi}}{sin\varphi}\right) = \frac{2 * 12}{18}\left(\frac{1}{0.30} - \frac{0.95}{0.37}\right) = 1.1 \text{ м}$$

$$E_a^x = H\sigma_{3c} = (8 - 1.1) * 31 = 214 \text{ кН}$$

For comparison, the active pressure and the maximum height of the vertical slope calculated by the equation [17] is:

$$\sigma_{3c} = \gamma H tg^2\left(45 - \frac{\varphi}{2}\right) + 2c * tg\left(45 - \frac{\varphi}{2}\right) = 65 - 16 = 48 \text{ кPa}$$

Height of steady slope

$$H = \frac{2c * cos\varphi}{\gamma(1 - sin\varphi)} = \frac{2 * 12 * 0.98}{15(1 - 0.2)} = 1.9 \text{ м}.$$

$$E_a^x = H\sigma_{\theta} = (8 - 1.9) * 48 = 292 \text{ кН}$$

3.5. The arguments of the physical nature of the coefficient of lateral pressure of the soil

As we know from classical mechanics, the coefficient of lateral earth pressure is a certain functional dependence that characterizes the ratio of the increments of the principal stresses in the absence of lateral movement, i.e. subject to the condition $\varepsilon_{2=3} = 0$

$$\xi = \frac{\Delta\sigma_3}{\Delta\sigma_1} \tag{3.11}$$

However, in practice, when solving engineering tasks, we are interested in the same ratio (4.29) only related to increasing deformations $0 < \varepsilon_3 \leq \varepsilon_u$, here ε_u – are the ultimate deformations of the soil. As shown by numerous experiments, $\xi = f(\varepsilon_u)$ has a decreasing character tending to a constant number (Figure 1.12, 1.24). The functional dependence of $\xi = f(\varepsilon_u)$ can be expressed as the following mathematical functions:

$$\xi = \xi_0 \left[(1/exp(k * \varepsilon_3)\right] \tag{3.12}$$

where k – correction factor; ξ_0 and ξ – can be characterized as parameters of the lateral soil pressure at rest (e.g. when exposed only to gravitational pressure in the absence of ground displacements) and transmitted to the ground external additional force impact when the possibility of lateral displacement. In accordance with the equation (3.12) for $\varepsilon_3 = 0$, $\xi = \xi_0$ and $\varepsilon_3 = \varepsilon_u$ we get the residual value of ξ. The nature of this phenomenon is associated with hardening of the soil in the process of moving soil particles on a surface. Therefore, the coefficient of lateral pressure is operatively associated with the main strength parameters of soil angle of internal friction. But its value is simpler and more reliable, for specific types of soils, determined in laboratory installations, for example, by testing on triaxial compression devices.

In Chapter 1, on the basis of triaxial soil, it was found that the lateral pressure and lateral extension of soil are closely related. Thus, when the condition $\sigma_1 \leq \sigma_{2=3}$, the deformation of soils corresponds to the condition of compression $\varepsilon_{2=3} = 0$. In this state of stress the value of the initial coefficient of lateral pressure corresponds to ξ_0 (Figure 1.24). When the condition $\sigma_1 > \sigma_{2=3}$ the value of this ratio begins to decline to the limit ξ_u. In this case, the following condition comes:

$$\sigma_1 > \sigma_{2=3} + k\mu \tag{3.13}$$

where k – coefficient for the studied sands $k \cong 440$.

For example, under extremely stressful conditions at $\mu > 0.4$

$$\sigma_u = 135 + 440 * 0.4 = 311 \ kPa$$

From the reference it is known [7,3,22] that the value of ξ_0 in the conditions of rest of the soil (for example, with the natural location of the soil layers) is usually determined by the formula (Zheka, 1944).

$$\xi_0 = (1 - sin\varphi) \tag{3.14}$$

In some cases, given the mismatch of equation (3.14) with the actual results of experiments, an additional correction factor k is introduced depending on the type of soil [22]:

$$\xi_0 = (1 - k * \sin\varphi) \tag{3.15}$$

where $\sin \varphi = (\sigma_3 - \sigma_1)/(\sigma_3 + \sigma_1)$.

As for the ultimate value ξ, it is determined by the results of triaxial soil tests, for example, in stabilometers. In accordance with the theory of strength of Coulomb-Mohr and Tresc(k)-Hill, it can be determined by the equations (2.13, 2.15).

It is believed that in the hydrostatic stress state when $\sigma_3 = \sigma_1$, the coefficient of lateral pressure is $\xi_0 = 1$. However, given the force of gravity, this value is always $\xi_0 < 1$.

We determine the physical meaning of the coefficient of lateral earth pressure as a medium having internal friction, using the model of Coulomb. In the environment we denote an area with an inclination angle $\theta = \varphi$. The plane strain on the inclined surface is the principal stresses and σ_3 σ_1. Moreover, the vertical stress σ_1 promotes mixing of the body obliquely downward and horizontal σ_3 keep the body in equilibrium. We expand the vertical and horizontal stresses in the normal and tangential τ_θ σ_θ voltage. We consider the equilibrium conditions for the site $\theta = \varphi$ (Figure. 1.19 a), 2.2).

We discussed the solution to this problem in Chapter 1. In accordance with the equation (1.19) in the limit equilibrium condition:

$$\frac{\tau}{\sigma} = \frac{tg\theta - \xi_N}{\xi_N tg\theta + 1} = tg\varphi \tag{3.16}$$

where $\theta = \varphi + \alpha_\xi$ deflection angle of the site. Equation (3.16) shows the relationship between the strength parameter φ, the angle of deviation of the most dangerous site of shear and the lateral pressure coefficient. We calculate equation (3.16) against the $k\xi_N$

$$\xi_N = \frac{tg\theta - tg\varphi}{1 + tg\theta tg\varphi} \tag{3.17}$$

or

$$tg\theta = \frac{\xi + tg\varphi}{1 - \xi * tg\varphi} \tag{3.18}$$

Thus, from equation (3.17) it follows that the coefficient of lateral pressure under the limit state of stress is a normal value, which characterizes the ratio of the external voltage and the internal resistance of the soil. From the equation (3.18) it follows that the angle of the maximum deviation of the shear pad of the soil depends not only on the magnitude of the angle of internal friction but also on the magnitude of the lateral pressure coefficient. For this angle, to solve practical problems of soil mechanics, we can take $\theta = \varphi + \alpha_\xi$.

3.5.1. Determination of the physical meaning of the coefficient of lateral pressure on the example of an infinite slope (plane problem)

We define the coefficient of lateral pressure corresponding to the condition of rest, i.e. when the relative deformation $\varepsilon = 0$. To solve this problem, we consider the stability of slopes *ABC* (Figure 3.8). For the angle of repose we take equal φ. Conventionally, from the slope we separate the persistent prism *AKM,* where $KM = h$ and $AM = MC = a$. Resistant *AKM* prism may be in a stable position only in a condition where the weight corresponds to the value

$$N_1 = \frac{ha}{2}\gamma = \frac{\gamma h^2}{2f} \tag{3.19}$$

where $a = \dfrac{h}{tg\varphi} = \dfrac{h}{f}$

In this case, the horizontal spacer forces of resistance acting on the persistent prism of the *AKM* are equal

$$T_1 = N_1 f = \frac{\gamma h^2}{2} \tag{3.20}$$

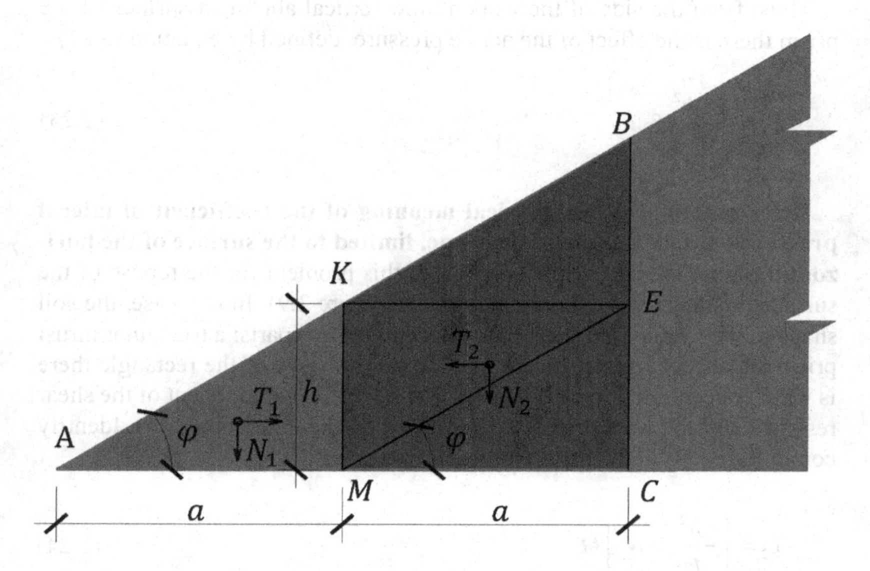

Figure 3.8 The analytical model of the slope to determine the coefficient of lateral pressure at rest, i.e. when the relative deformation ε = 0.

On the other hand, the gravitational forces from the *MKBC* slope act on the persistent prism of the *AKM*. It is not difficult to determine what corresponds to its weight:

$$N_2 = 3N_1 = 3\frac{\gamma h^2}{2f}$$

$$\text{and } T_2 = \xi N_2 = 3\frac{\gamma h^2}{2f}\xi \tag{3.21}$$

The stable position of the slope should satisfy $T_1 > T_2$, which implies that

$$f > 3\xi \text{ or } \xi_{\varepsilon=0} < \frac{f}{3} \tag{3.22}$$

where $f = tg\varphi$, φ – the angle of repose.

Thus, from the side of the slope to the vertical abutment surface of the prism there is the effect of the active pressure, defined by equation (3.21)

$$E_a = \frac{3\gamma h^2}{2f}\xi \tag{3.23}$$

Determination of the physical meaning of the coefficient of lateral pressure on the example of the slope, limited to the surface of the horizontal plane. We regard the solution of this problem for the repose of the surface bounded by the horizontal plane (Figure 3.9). In this case, the soil shape in the form of a trapezoid is divided into two parts: a triangular thrust prism and a rectangular. Imagine that on the surface of the rectangle there is valid concentrated force $N = q*a$. The horizontal component of the shear resistance of the hard prism is determined by the equation (3.20). Identify common spacers' effort from a rectangular prism:

$$T_2 = \left(\frac{\gamma h^2}{f} + N\right)*\xi \tag{3.24}$$

When $N = N_u$ the slope will be in an extreme state of equilibrium. To this end, the equation (3.20) equates to (3.24) and solving the equation for N_u we get

$$N_u = \frac{\gamma h^2}{\xi}\left(\frac{1}{2} - \frac{\xi}{f}\right) = \frac{\gamma h^2}{2f}\left(\frac{f}{\xi} - 2\right) \tag{3.25}$$

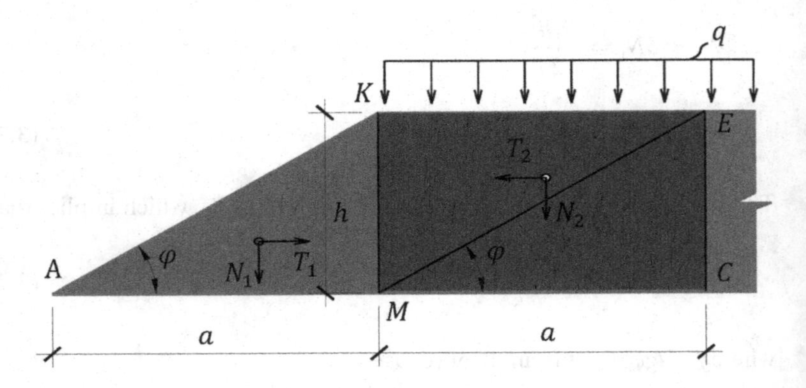

Figure 3.9 The analytical model of the slope to determine the coefficient of lateral pressure at rest, i.e. when the relative deformation $\varepsilon = 0$ and the critical load N on a horizontal ground surface.

Here $\xi = \xi_u$ – corresponds to the critical value of the lateral pressure coefficient. The magnitude of the force N_u at quiescence condition, i.e. when $\xi = \xi_{\varepsilon=0}$ (3.25) is equal to the weight of the cut part of the upper slope from the slope body (case of initial critical load)

$$N_{Ru} \approx \frac{\gamma h^2}{2f} \tag{3.26}$$

$$\left(\frac{f}{\xi} - 2\right) = 1$$

$$\xi_{\varepsilon=0} < \frac{f}{3} \tag{3.27}$$

Thus, for this case, the slope, the equation (3.27) is consistent with (3.22).

3.6. Calculation of resistance and ultimate load on the base

It is known that the bearing capacity of the base under the loaded surface is calculated on the basis of two principal positions:

- The Terzaghi method [15], based on the appearance of a shear limit surface along a compacted wedge under the foundation. It is believed that the shear surface AB is deflected from the base of the foundation by an angle φ
- The method of complete loss of stability [3, 7, 17]. In this case, solid shear surfaces appear at the base, provoking the stability of the base.

According to the first method, the maximum load on the soil from the distributed belt, square and round external load is determined by the equations:

$$q_u = c^1 N_c^1 + qN_q^1 + 0.5 * \gamma(2b) N_\gamma^1$$

$$q_u = 1.3c^1 N_c^1 + qN_q^1 + 0.4 * \gamma(2b) N_\gamma^1 \tag{3.28}$$

$$q_u = 1.3c^1 N_c^1 + qN_q^1 + 0.3 * \gamma(2b) N_\gamma^1$$

Where: $c^1 = \frac{2}{3}c$ and $\varphi^1 = \frac{2}{3}\varphi$; N_c^1, N_q^1 and N_γ^1 coefficients determined depending on the angle of internal friction. In case of $\varphi^1 = 0$ и $\tau = c^1$; $N_\gamma^1 = 5,7$; $N_q^1 = 1$; $N_c^1 = 5,7$.

Krizek – proposed these coefficients of equations (1) are determined by the equations:

$$N_c^1 = \frac{228 + 4,3\varphi^1}{40 - \varphi^1}; \; N_q^1 = \frac{40 + 5\varphi^1}{40 - \varphi^1} \text{ and } N_\gamma^1 = \frac{6\varphi^1}{40 - \varphi^1} \tag{3.29}$$

The authors propose to determine these coefficients by equations

$$N_c^1 = 1,8 * e^{\left(0,1 * \varphi^1\right)}; \; N_q^1 = 0,3 * e^{\left(0,14 * \varphi^1\right)} \text{ and}$$

$$N_\gamma^1 = 0,35 * e^{\left(0,14 * \varphi^1\right)} \tag{3.29.1}$$

According to the second method, when the shear surfaces cover the entire base, the ultimate load (1) on the ground from the distributed:

$$q_u = q_C + q_q + q_\gamma$$

$$q_u = c^1 N_c^1 + q N_q^1 + 0.5 * \gamma(2b) N_\gamma^1 \tag{3.30}$$

Where, N_c^1, $N_q^1 и N_\gamma^1$ external load is determined with the coefficients [2]:

$$N_c^1 = \left(N_q^1 - 1\right) * cot * \varphi^1); \; N_q^1 = tan^2\left(45 + \frac{\varphi^1}{2}\right) * e^{\left(\pi * tan\varphi^1\right)}$$

and

$$N_\gamma^1 = \left(N_q^1 - 1\right) * tan * (1.4 * \varphi^1); \tag{3.31}$$

The solution of this task by L Prandtl and G. Reisner (1920–1921) was obtained as the following equation:

$$q_u = (\gamma d - c\,cot\varphi)\frac{1 + sin\varphi}{1 - sin\varphi} e^{\pi tg\varphi} - c\,cot\varphi \tag{3.32}$$

Based on the studies [5], we assume that the base is affected by an average stress equal to the limiting state q_u. On the sides there is a load on the weight of the soil layer equal to q_2. The amount of natural pressure on the surface BC q_1

$$q_1 = q_g = b\gamma tg\varphi^*$$

$$N_g = q_1 * \text{BC} = \gamma tg\varphi^* * \frac{b^2}{cos^2\varphi^*}$$

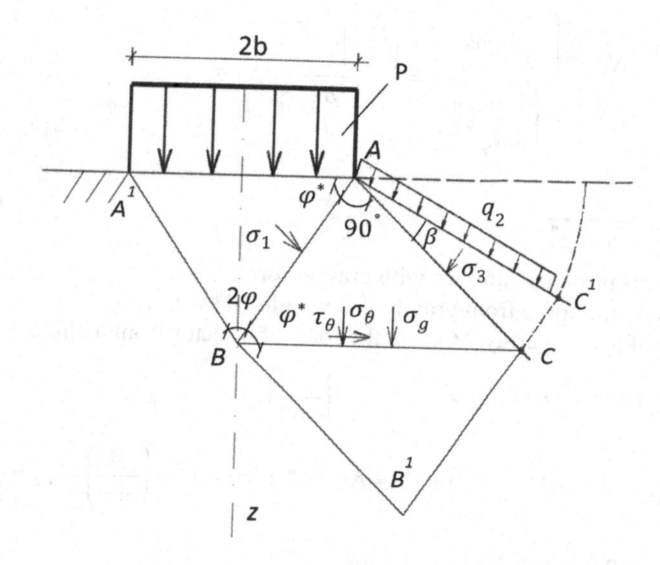

Figure 3.10 Design scheme for determining the critical load on the ground

The stability of the compacted wedge under the baseline ABA^I is ensured both by the strength of the resistant oblique formation ($I\text{-}ABB^IC$) on the one hand and by the soil load ACC^I (II).

In this task, in contrast to the above, the strength condition of an inclined rectangular prism ($I\text{-}ABB^IC$) is taken as the stability criterion (Figure 3.1). It is assumed that the main stresses act on the verge of a prism: σ_1 – from the side of the compacted wedge AB and σ_3 – from the inclined profile AC. The side of an inclined rectangular prism is connected to the surface AB. Distributed load acts on inclined surface (load) q_2. Determine the condition of strength and stability resistant oblique reservoir (I-ABC). It is assumed that the shift will occur at the site BC (Figure 3.10) deflected from the main site at an angle $\varphi^* = (\varphi + atan(\xi))$. Where ξ is the lateral pressure ratio. As our studies have shown, the values of tangential stresses arising along the base of the foundation on this surface decrease to zero. The design scheme is presented in Figure 3.10.

Determination of the ratio of the forces acting on the surface of the compacted wedge. We assume that the main areas of the AB and AC are forces N_1 and N_3.

$$N_1 = \left(\frac{N_u}{2}\right) cos\varphi^* \tag{3.33}$$

$$\sigma_1 = \frac{N_1}{AB} = \frac{\left(\frac{N_u}{2}\right) cos\varphi^*}{\left(\frac{b}{cos\varphi^*}\right)} = \frac{\left(\frac{N_u}{2}\right) cos^2\varphi^*}{b} = \sigma_u cos^2\varphi^*$$

$$\sigma_u = \frac{\sigma_1^*}{cos^2\varphi^*} \tag{3.34}$$

Where σ_1^*–principal stresses with gravity forces.
Determine the force from gravity and weight (II)ACC^1.
Weight of leaning array ACC^1 in the form of a sector from a circle

$$N_g\left(AC * CC^1\right) = 2 * \pi * AC^2 \gamma \left(\frac{\beta}{360}\right)$$

$$N_3 = N_g\left(AC * CC^1\right) + q_2 * AC = 2 * \pi * AC^2 \gamma \left(\frac{\beta}{360}\right) + q_2 * \frac{btg\varphi^*}{cos\varphi^*}$$

$$\sigma_3 = 2 * \pi * AC^2 \gamma \left(\frac{\beta}{360}\right) + q_2 \tag{3.35}$$

Make an equation of equilibrium of all forces on the surface BC and subordinate it with the shift condition equation

$$N_3 cos\varphi^* + N_1 \sin\varphi^* = tg\varphi\left(N_3 \sin\varphi^* + N_1 cos\varphi^* + N_g(BC)\right)$$

$$N_1\left(\sin\varphi^* - cos\varphi^* tg\varphi\right) = N_3\left(\sin\varphi^* tg\varphi - cos\varphi^*\right) + N_g(BC) tg\varphi$$

$$N_1 = N_3 \frac{\left(\sin\varphi^* tg\varphi + cos\varphi^*\right)}{\left(\sin\varphi^* - cos\varphi^* tg\varphi\right)} + \frac{N_g(BC)}{\left(\sin\varphi^* / tg\varphi - cos\varphi^*\right)}$$

$$\sigma_1 = \frac{N_1}{AB} = N_3 \frac{\left(\sin\varphi^* tg\varphi + cos\varphi^*\right)}{\left(\sin\varphi^* - cos\varphi^* tg\varphi\right)} \left(\frac{cos\varphi^*}{b}\right) + \frac{N_g(BC)}{\left(\sin\varphi^* / tg\varphi - cos\varphi^*\right)} \left(\frac{cos\varphi^*}{b}\right)$$

$$\sigma_1 = \frac{N_1}{AB} = N_3 \frac{\left(tg\varphi^* tg\varphi + 1\right)}{\left(tg\varphi^* - tg\varphi\right)} \left(\frac{cos\varphi^*}{b}\right) + \frac{N_g(BC)}{\left(\sin\varphi^* / tg\varphi - cos\varphi^*\right)} \left(\frac{cos\varphi^*}{b}\right)$$

$$\sigma_1 = \frac{N_1}{AB} = N_3 \frac{\left(tg\varphi^* tg\varphi + 1\right)}{\left(tg\varphi^* - tg\varphi\right)} \left(\frac{tg\varphi^*}{\frac{tg\varphi^* b}{cos\varphi^*}}\right) + \frac{N_g(BC)}{\left(\sin\varphi^* / tg\varphi - cos\varphi^*\right)} \left(\frac{cos\varphi^*}{b}\right)$$

$$\sigma_1 = \frac{N_1}{AB} = \sigma_3 \frac{\left(tg\varphi^* tg\varphi + 1\right)}{\left(tg\varphi^* - tg\varphi\right)} tg\varphi^* + \frac{cos\varphi^* N_g\left(BC\right)}{cos\varphi^*\left(\sin\varphi^*/tg\varphi - cos\varphi^*\right)}\left(\frac{cos\varphi^*}{b}\right)$$

$$\sigma_1 = \frac{N_1}{AB} = \sigma_3 \frac{\left(tg\varphi^* tg\varphi + 1\right)}{\left(tg\varphi^* - tg\varphi\right)} tg\varphi^* + \frac{\sigma_g\left(BC\right)}{cos\varphi^*\left(\sin\varphi^*/tg\varphi - cos\varphi^*\right)} \quad (3.36)$$

Denote

$$M_\gamma = \frac{\left(tg\varphi^* tg\varphi + 1\right)}{\left(tg\varphi^* - tg\varphi\right)} tg\varphi^*;\ M_g = \frac{1}{cos\varphi^*\left(\sin\varphi^*/tg\varphi - cos\varphi^*\right)} \quad (3.37)$$

$$\sigma_1 = M_\gamma \sigma_3 + M_g \sigma_g\left(BC\right) = M_\gamma \sigma_3 + M_g b\gamma tg\varphi^* \quad (3.38)$$

The maximum load on the soil will be determined by the equation (3.33)

The task is more simply solved using the theory of strength proposed by the authors. In accordance with the equation (2.21) we define:

$$\sigma_1^* = \sigma_3/\xi$$

$$P = \sqrt{\sigma_1^2 l^2 + \sigma_3^2 s^2} = \sigma_3 \sqrt{\left(\frac{1}{\xi}\right)^2 l^2 + s^2} \quad (3.39)$$

$$\sigma_\theta = \mathrm{P}\cos\theta + q_1 \quad (3.40)$$

$$\tau_\theta = \mathrm{P}\sin\theta$$

In conditions of extreme stress state $\theta = \varphi$. In this case

$$\sigma_1^* \cong \frac{\sigma_3}{\xi} + 2{,}5q_1 = \frac{\sigma_3}{\xi} + 2{,}5b\gamma tg\varphi^* \quad (3.41)$$

Finally, the maximum load on the soil will be determined by the equation (3.34)

$$\sigma_u = \frac{\sigma_1^*}{cos^2\varphi^*} = \frac{1}{cos^2\varphi^*}\left(\frac{2\pi\gamma\left(\frac{btg\varphi^*}{cos\varphi^*}\right)^2\left(\frac{\beta}{360}\right) + q_2}{\xi} + 2{,}5b\gamma tg\varphi^*\right) \quad (3.42)$$

Task

It is required to determine the marginal carrying capacity of the base σ_u. Initial data $b = 2m$. soil density $\gamma = 16\,kN/m^2$, Angle of internal friction $\varphi = 38^0$. The lateral pressure coefficient according to the results of triaxial tests

$\xi = 0,28$. Shear surface deflection angle $\varphi^* = [\varphi + atan(\xi)] = 38 + 15 = 53^0$. $c^1 = 0$. Foundation depth d = 0,5 m. $q = \gamma * \text{d} = 16 * 0,5 = 8$ kN / m. Calculating according to the equation (3.28)

$$q_u = c^1 N_c^1 + qN_q^1 + 0.5 * \gamma(2b)N_\gamma^1 = 8 * 61,55 + 0,5 * 16 * 4 * 78,61$$
$$= 3008\ k\text{Pa}$$

According to the second method (3.30), the maximum load on the soil from the distributed, external load is determined by the equation:

$$q_u = c^1 N_c^1 + qN_q^1 + 0.5 * \gamma(2b)N_\gamma^1$$
$$q_u = 8 * 48,8 + 0.5 * 16(2 * 2) * 64 = 2438 k\text{Pa}$$

By the third method (3.32) $q_u = 391 k\text{Pa}$

The calculation of the equation (3.28), taking into account the tabular coefficients, determined by the equations (3.28) and (3.29). The results are shown in table 3.5.

According to the method proposed by the authors, the determination of the ultimate loads on the base depends on the slope β. The results of the calculations are presented in Table 3.6.

As follows from the results of the calculation, the ultimate load, unlike the above, among other things, depends both on the angle of inclination β and the value of the coefficient of lateral pressure. For a horizontal surface slope $\beta = 38^0$ $q_u = 3138 kPa$.

For $\varphi = 20^0$ and $\xi = 0,22$ (other indicators remain unchanged): for the first task, respectively: q_u = 132; 223; and 176 kPa. By the second method $q_u = 143 k\text{Pa}$. By the third method (3.32) $q_u = 52 k\text{Pa}$ and for the method proposed by the authors when $\beta = 38^0$ $q_u = 225 k\text{Pa}$.

Table 3.5 The results of the calculation of the ultimate loads on the grounds

Coefficients	*Tabular*	*By equations (3.29)*	*By equations (3.29.1)*
N_c^1	77,50	27,33	71,5
N_q^1	61,55	14,22	61,3
N_γ^1	78,61	13,22	80,5
q_u (kPa)	3008	536	2780

Table 3.6 The results of the calculation of the ultimate loads on the base determined by the equations (3.38, 3.42)

β	φ	ξ	φ^*	b	γ	q_2	D	σ_3	σ_1^*	σ_q	M_γ	M_g	σ_1	q_u
°	°		°	*m*	*kH*/m³*	*kH/m*	*m*		*kPa*				*kPa*	
0,0	38,0	0,3	53,6	2,0	16,0	8,0	0,5	8,0	137,2	43,5	4,9	3,9	50,1	142
2,0	38,0	0,3	53,6	2,0	16,0	8,0	0,5	19,7	179,1	43,5	4,9	3,9	107,0	304
4,0	38,0	0,3	53,6	2,0	16,0	8,0	0,5	31,5	221,0	43,5	4,9	3,9	163,9	466
6,0	38,0	0,3	53,6	2,0	16,0	8,0	0,5	43,2	262,9	43,5	4,9	3,9	220,8	628
8,0	38,0	0,3	53,6	2,0	16,0	8,0	0,5	54,9	304,8	43,5	4,9	3,9	277,8	790
10,0	38,0	0,3	53,6	2,0	16,0	8,0	0,5	66,7	346,7	43,5	4,9	3,9	334,7	952
12,0	38,0	0,3	53,6	2,0	16,0	8,0	0,5	78,4	388,6	43,5	4,9	3,9	391,6	1114
14,0	38,0	0,3	53,6	2,0	16,0	8,0	0,5	90,1	430,5	43,5	4,9	3,9	448,5	1276
16,0	38,0	0,3	53,6	2,0	16,0	8,0	0,5	101,8	472,4	43,5	4,9	3,9	505,4	1438
18,0	38,0	0,3	53,6	2,0	16,0	8,0	0,5	113,6	514,3	43,5	4,9	3,9	562,3	1600
20,0	38,0	0,3	53,6	2,0	16,0	8,0	0,5	125,3	556,2	43,5	4,9	3,9	619,2	1762
22,0	38,0	0,3	53,6	2,0	16,0	8,0	0,5	137,0	598,1	43,5	4,9	3,9	676,2	1923
24,0	38,0	0,3	53,6	2,0	16,0	8,0	0,5	148,8	640,0	43,5	4,9	3,9	733,1	2085
26,0	38,0	0,3	53,6	2,0	16,0	8,0	0,5	160,5	681,9	43,5	4,9	3,9	790,0	2247
28,0	38,0	0,3	53,6	2,0	16,0	8,0	0,5	172,2	723,8	43,5	4,9	3,9	846,9	2409
30,0	38,0	0,3	53,6	2,0	16,0	8,0	0,5	184,0	765,7	43,5	4,9	3,9	903,8	2571
37,0	38,0	0,3	53,6	2,0	16,0	8,0	0,5	225,0	912,3	43,5	4,9	3,9	1103,0	3138

Calculation of stability and the maximum inclined (horizontal) load on the base. The condition of the bearing capacity of the base is determined by the equation:

$$F \leq \gamma_c F_u / \gamma_n \quad (3.43)$$

Where F – resultant design load on the base under the action F_v and F_h. inclined to the vertical angle δ; F_u – strength of limiting resistance. γ_c and γ_n – working conditions coefficients $\gamma_c = (1-0,9)$ and foundation reliability $\gamma_n = (1,2-1,1)$. The maximum inclined load on the base is determined by

$$F_u = \sqrt{F_v^2 + F_h^2} = F_v\sqrt{1+(tg\delta)^2}$$

Studies have shown that the limiting deviation angle can be taken equal to $\delta_u = 0,65\varphi$. For example, as shown die tray (D = 15cm; A = 180 cm²) studies

Figure 3.11 General view of the installation to determine the ultimate load on the soil base

based on coarse sands on vertical and horizontal loads; a complete breakdown of the stamp (sands are glued on the bottom of the stamp) occurred at a ratio of $F_h / F_v = 0,46$. Experiments with stamps loaded with limiting vertical and horizontal loads shown in Figure 3.2. Depth of the stamp d = 0. Wherein $\delta = 26^0$. The limiting angle of inclination of the resultant force is obtained on the basis of the experiments performed and can be determined by the equation

$$\frac{\tau_{xz}}{\sigma_1} = tg\delta \cong 0,65 * tg\varphi \tag{3.44}$$

Charts of depending of stress σ_m, τ_m and moving u_m, s_m stamp; and horizontal moving of soil $(x = -r)$ at depth $\frac{z}{d}$ is shown on Figures 3.13,

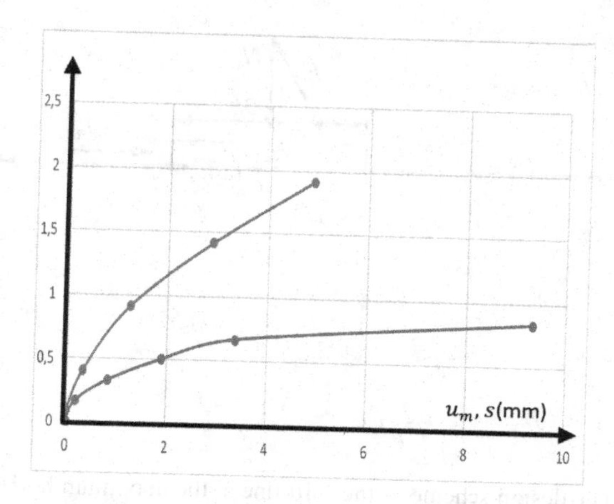

Figure 3.12 Dependence between strength and movements. 1-at normal and 2 at tangential stresses

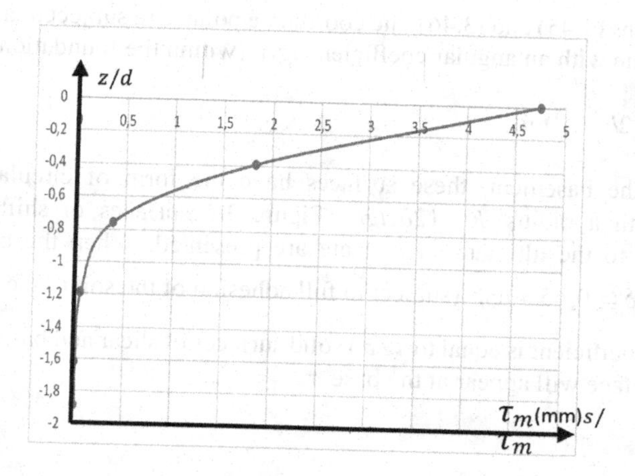

Figure 3.13 Dependence of changes in horizontal movements in depth.

3.14. It has been established that the function of horizontal displacements and stresses agrees well with the equations (3.45, 3.46) (Figure 3.14).

$$s_x = \frac{\tau_m}{G} e^{\left(-\frac{\pi(x-2b)z}{d}\right)} \tag{3.45}$$

$$\tau_{xz} = \tau_m e^{(-2.2 * tg\alpha)} \tag{3.46}$$

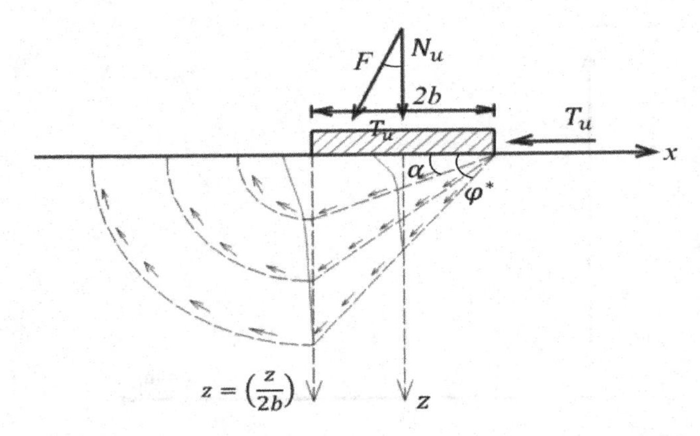

Figure 3.14 The design scheme of the shift line at the maximum horizontal (slope) strength

In equations (3.45) and (3.46), the coordinate points are subject to a straight oblique line with an angular coefficient $tg\alpha$ (within the foundation).

$$z = (2b - x)tg\alpha \tag{3.47}$$

Outside the basement, these surfaces have the form of circular cylindrical with a radius $R = (2b)tg\alpha$. Figure 3.5 surfaces of shifts corresponding to the ultimate stress state are presented. When the condition $\frac{\tau_{xz}}{\sigma_1} = tg\delta \cong 0,65 * tg\varphi$ (subject to full adhesion of the soil to the sole, the angular coefficient is equal to $tg\varphi$) solid surfaces of shear and base sticking to the surface will appear at the base $\tau_{xz} = \tau_{uxz}$.

Task

Required to determine the marginal carrying capacity of the base τ_u and surface shear on an oblique load. The vertical component of the average strength $\sigma_m = 200$ kPa. Initial data $2b = 2$m. Angle of internal friction $\varphi = 38^0$. Define the maximum deflection angle $\varphi^* = [\varphi + atan(\xi)] = 38 + 15 = 53^0$. $c^1 = 0$. Foundation depth d = 0m. Calculation by equation (1)

$$\tau_{uxz} = \sigma_m tg\delta \cong 200 * 0,65 * tg\varphi = 101\text{kPa}$$

The calculation of the values of the limit tangential stresses satisfying the Coulomb strength condition is determined by equation (3.29).

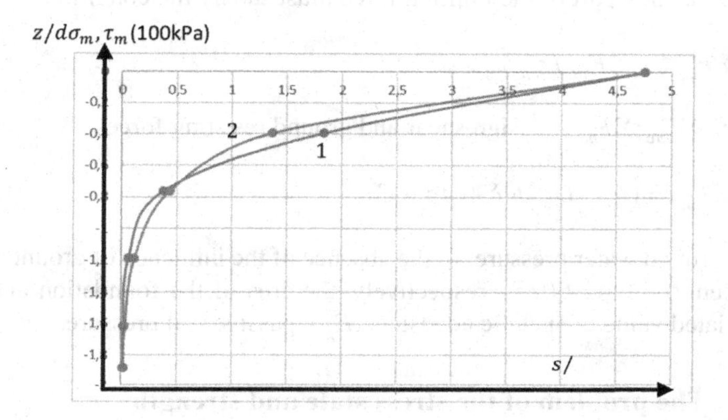

Figure 3.15 Dependence between strength and displacement.1-experimental and 2 theoretical.

Table 3.7 The results of the calculation of limiting shear strengths

σ_m	$0 \le \delta \le 0.5\varphi$	τ_m	$2{*}b$	x	z	$0 \le \alpha \le \varphi^*$	τ_{ui}
200	26,794	101	2	0	0,00	0	101,00
200	26,794	101	2	0	0,35	10	68,53
200	26,794	101	2	0	0,73	20	45,35
200	26,794	101	2	0	1,15	30	28,36
200	26,794	101	2	0	1,68	40	15,94
200	26,794	101	2	0	2,38	50	7,34

The results are shown in the Table 3.7

The lower limit of the shear limit surface, where the tangential displacements are completely absent is determined by the equation (3.47) and corresponds to the inclined line with coordinates:

$$z_1 = (2b - x)tg\varphi^* = (2 - 0)tg53 = 2,65 \text{ м. и}$$
$$z_2 = (2b - x)tg\varphi^* = (2 - 2)tg53 = 0 \text{ м}$$

The final inclined force on the foundation is determined by the equation

$$F_u = \sqrt{F_v^2 + F_h^2} = F_v\sqrt{1 + (tg\delta)^2} = F_v\sqrt{1 + (0,65)^2} = 1,2F_v$$

The horizontal part of the limiting force must satisfy the condition

$$\Sigma F_h \leq \gamma_c \Sigma F_{hu} / \gamma_n \tag{3.48}$$

Where $\Sigma F_{hc}; \Sigma F_{hu}$ – design shear and ground resisting forces.

$$\Sigma F_{hu} = \left(F_v - U_w\right) tg\delta + Ac + E_p \tag{3.49}$$

U_w – ground water pressure. In the absence of the influence of groundwater is taken $U_w = 0$; A, c – respectively, the area of the foundation and the calculated value of specific adhesion; E_p – passive soil pressure.

3.7. The problem of the stress state and strength of the borehole walls with axisymmetric radial effective stresses

Theoretical solutions of the asymmetric problem of a thick-walled pipe are known [1]. In practice, this problem can be used as a basis for solving the axisymmetric problem in geotechnics. In practice, this task can be applied in determining the critical load on the walls of the well or in determining the strengthen deformed condition (SDC) around it in the presence of gravitational pressure of the soil mass. This raises the fair question of tensile strengths and deformation in the soil mass in the tangential direction. The fact is that, unlike the elastic problem of a thick-walled pipe, strengths and strains in the tangential direction are absent or excessively small due to the limited tensile strength of the soil. Despite these limitations, to solve the geotechnical problem of wells with internal pressure, we will take the Lame [1] solution as a basis.

Briefly look at Lame's elastic solution. Denote by ε_r – radial deformation (in the direction of the radius of the cylinder), ε_t – circumferential or tangential deformation. Linear deformations and normal strengths occur in three mutually perpendicular directions: $\sigma_r = E\varepsilon_r$ and $\sigma_t = E\varepsilon_t$. If in the axial direction along the hole the deformation is free, i.e. there are no external constraints that prevent elastic changes in the length of the pipe $\sigma_x = 0$;

Consider the effect of two pressures: internal pressure p_1 and external p_2, evenly distributed on the inner and outer surfaces of the hollow cylinder. Its external and internal radii are denoted, respectively, by r_0 and r_1. Let us determine the forces transmitted at a specified load to an arbitrarily chosen elementary volume dV (Figure 3.16a,b).

We compose the equation of equilibrium of forces in the radial direction along the axis R

$$(\sigma_r + d\sigma_r)(r + dr)d\theta \cdot 1 - \sigma_r r d\theta \cdot 1 - 2\sigma_t \cdot 1 \cdot sin\frac{d\theta}{2} = 0$$

Neglecting infinitesimal quantities of the second order of smallness and taking

$sin\frac{d\theta}{2} \approx \frac{d\theta}{2}$, we receive $\frac{d\sigma_r}{dr} + \frac{\sigma_r - \sigma_t}{r} = 0.$ (3.50)

at the case under consideration, differential equation (1) is the only one obtained from the equilibrium condition, but with two unknowns σ_r и σ_t. Therefore, the task will be statically indefinable.

In relation to soils, taking into account the theory of strength, it can be argued that the tangential stresses must satisfy the condition $0 \le \sigma_t \le c\,ctg\varphi$. Where c ,$\varphi$ – accordingly, the specific adhesion and the angle of internal friction. With this in mind, i.e. $\sigma_t = 0$ differential equation (1) becomes homogeneous and has the form

$$\frac{d\sigma_r}{dr} + \frac{\sigma_r}{r} = 0. \tag{3.51}$$

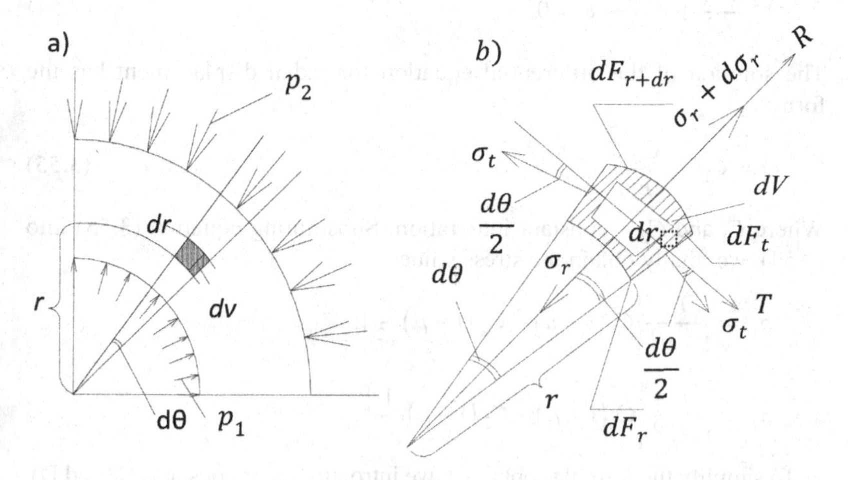

Figure 3.16 Design schemes: a) a rectangular segment in plan and b strengthen condition in an elementary volume

In the basic version, the differential equation has two unknowns σ_r *and* σ_t. In order to reduce the Task (3.50) to a uniform one, use Hooke's law. The second equation – the equation in displacements:

$$\varepsilon_r = \frac{1}{E}(\sigma_r - \mu\sigma_t) = \frac{du}{dr}$$

$$\varepsilon_t = \frac{1}{E}(\sigma_t - \mu\sigma_r) = \frac{du}{dr} \tag{3.52}$$

Solving these two equations with respect to σ_r and σ_t, can get equation in displacements

$$\sigma_r = \frac{E}{1-\mu^2}\left(\frac{du}{dr} + \mu\frac{u}{r}\right);$$

$$\sigma_t = \frac{E}{1-\mu^2}\left(\frac{u}{r} + \mu\frac{du}{dr}\right) \tag{3.53}$$

In equations (3.53), both voltages σ_r and σ_t expressed through a variable u, i.e. through the magnitude of the radial displacement corresponding to the radius r.

Substituting the strength values (3.53) into equation (3.51), we obtain a differential equation expressed in displacements

$$r^2\frac{d^2u}{dr^2} + r\frac{du}{dr} - u = 0. \tag{3.54}$$

The solution of the differential equation for radial displacement has the form:

$$u = C_1 r + \frac{C_2}{r}. \tag{3.55}$$

Where C_1 and C_2 – constant integration. Substituting equation (3.55) into (3.54), we finally obtain the stress values

$$\sigma_r = \frac{E}{1-\mu^2}\left[C_1(1+\mu) + C_2(1-\mu)\frac{1}{r^2}\right];$$

$$\sigma_t = \frac{E}{1-\mu^2}\left[C_1(1+\mu) - C_2(1-\mu)\frac{1}{r^2}\right].$$

To simplify the formulas obtained, we introduce other constants (C and D), giving the last two equations the following form

$$\sigma_r = C + \frac{D}{r^2}; \tag{3.56}$$

Figure 3.17 General view of the test bench.

$$\sigma_t = C - \frac{D}{r^2};$$

It is not difficult to notice that equation (6) has a similar form obtained for the plane problem [4]. This equation allows one to find both radial and circumferential tangential stresses for an arbitrary value of an independent variable r, i.e., for any concentric soil layer in the well. The values of the constants C and D are determined from the boundary conditions.

Solving the system of these equations for C and D and accepting the condition $r_1 \cong 6r_0$ it is easy to get that

$$C = \frac{p_1 r_0^2 - p_2 (6r_0)^2}{(6r_0)^2 - r_0^2} = \frac{p_1 - 36p_2}{35};$$

$$D = (p_2 - p_1)\frac{(6r_0)^2 r_0^2}{r^2 - r_0^2} = (p_2 - p_1)\frac{36r_0^2}{35} \cong (p_2 - p_1)r_0^2$$

Consider the case when $p_1 > p_2$. In this case, the radial strengths are positive, i.e. soil in this direction works for compression $\sigma_r > 0$. Negative tangential strengths $\sigma_t < 0$. Recall that the quantity p_1 – in relation to soils, it is determined from the condition that there is no movement at a distance H_s equal to the active region of compression of the concentric layer determined on the basis of experiments $\varepsilon_r = 0$. In the Figures 3.15 and 3.17 presents an experimental graph of the dependence of changes in radial displacements $(r_0 + r)/d$ from movements (u/p_2).

As can be seen from the experimental results for sands of medium size, the active compression region is - $H_s = 3d = 6r_0$.

Horizontal external gravitational stresses p_2 can be determined from the condition of equality:

$$p_2 = \sigma_{zgx} = \xi\gamma_{zg} z \tag{3.57}$$

Where γ_{zg}, z and ξ – accordingly, the weighted average specific gravity of the soil, the depth of the layer and the coefficient of lateral pressure of the soil.

Subject to conditions $\sigma_t \cong 0$ consider the solution of the homogeneous differential equation (3.51). We transform equation (3.51) to the form

$$\frac{d\sigma_r}{\sigma_r} + \frac{dr}{r} = 0. \tag{3.58}$$

the solution of the homogeneous differential equation (3.58) is written as follows

$$ln\sigma_r = ln\left(\frac{C}{r}\right) \text{ or } \sigma_r = \frac{C}{r} \tag{3.59}$$

Where C – the integration constant is determined from the boundary conditions for $r = r_0$ $\sigma_r = p_2$ from where $p_2 = \frac{C}{r_0}$ or $C = p_2 r_0$. And so we finally write equation (3.59) in the form

$$\sigma_r = p_2 \frac{r_0}{r} \tag{3.60}$$

As shown by the results of tray experiments (Figure 3.17), radial stresses in a soil medium σ_r decrease more intensively and at $\left(\frac{r_0}{r}\right) = \left(\frac{r_0}{6r_0}\right) \cong 0$. the condition is satisfied $\sigma_r \cong 0$. Therefore the equation (3.60) for practical purposes, accuracy as applied to soils can be written as a power function of the form

$$\sigma_r = p_2\left(\frac{r_0}{r}\right)^k, \tag{3.61}$$

where k – coefficient determined experimentally and depending on the type of soil varies in the aisles $k = (1{,}5 - 2{,}5)$.

The coordinate function of the form also satisfactorily meets these requirements.

$$\sigma_r = p_1\left(-0{,}0184\left(\frac{r}{r_0}\right)^3 + 0{,}22\left(\frac{r}{r_0}\right)^2 - 1{,}1946\left(\frac{r}{r_0}\right) + 1.9477\right) \tag{3.62}$$

Or in a more compact form

$$\sigma_r = p_1\exp\left[-k\left(r - r_0\right)\right] \tag{3.63}$$

where $k = (1 - 1{,}5)$ – experimentally determined coefficient.

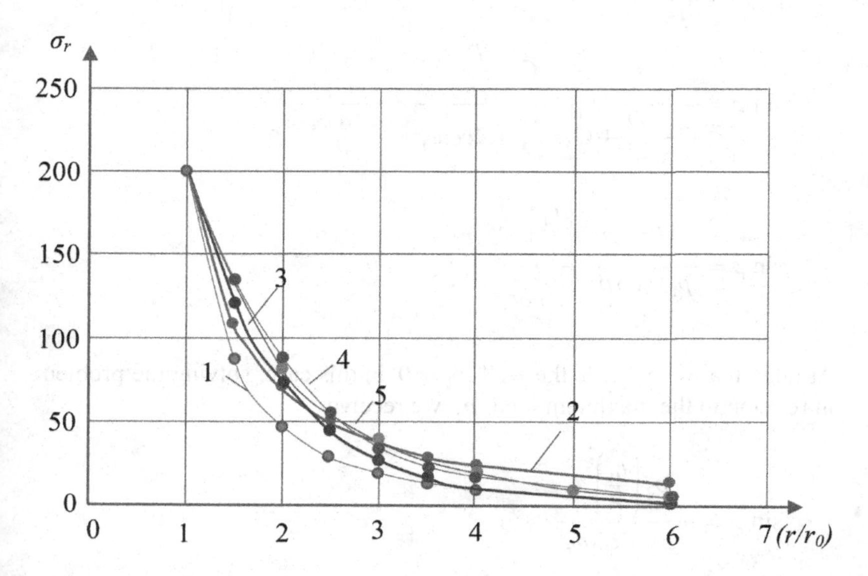

Figure 3.18 Dependences of the change in radial stresses on the reduced distance (formula). 1,2,3,4 – appropriately calculated by equations 7, 12, 14, 13 and 5 – results of tests.

For example, for equations (3.61) for $r = 6r_0$ $(\sigma_r - \sigma_{zg}) = 0,07p_1$; (3.62) $(\sigma_r - \sigma_{zg}) = p_2\left(\frac{1}{6}\right)^2 = 0.028 * p_2$; for equation (3.63) $(\sigma_r - \sigma_{zg}) = 0,007p_1$ and for equation (3.56) $(\sigma_r - \sigma_{zg}) = 0,11p_1$

The calculated results were compared with the experimental displacement curve with a maximum function $\sigma_{rmax} = f(u) = 1$

As can be seen from the graph, close to the experimental results correspond to coordinate functions of the form (12-13-14).

Consider the case when $p_2 > p_1$. In this case, both radial and tangential stresses are positive, i.e. soil in this direction works for compression. If these strengths are taken as the main ones, then in accordance with the Coulomb-Mohr theory of strength, they must satisfy the condition

$$\sin\varphi = \frac{\sigma_t - \sigma_r}{\sigma_t + \sigma_r + 2c\cot\varphi} \tag{3.64}$$

Substituting the values (3.56) into the hostility (3.64), we obtain

$$D = (p_2 - p_1)r_0^2$$

$$C = \frac{p_1 - 36p_2}{35}$$

$$\sin\varphi = \frac{C + \frac{D}{r^2} - C + \frac{D}{r^2}}{C - \frac{D}{r^2} + C + \frac{D}{r^2} + 2c\cot\varphi} = \frac{\frac{D}{r^2}}{C + c\cot\varphi}$$

$$\sin\varphi = \frac{(p_2 - p_1)\left(\frac{r_0}{r}\right)^2}{\frac{p_1 - 36p_2}{35} + c\cot\varphi}$$

Assume that with inside the well $p_1 = 0$ in this case, solving the problem in relation to the maximum load p_2 we receive

$$\sin\varphi = \frac{p_2\left(\frac{r_0}{r}\right)^2}{-p_2 + c\cot\varphi}$$

$$(c\cos\varphi) = p_2\left[\sin\varphi + \left(\frac{r_0}{r}\right)^2\right]$$

$$p_2 = \frac{c\cos\varphi}{\left[\sin\varphi + \left(\frac{r_0}{r}\right)^2\right]} \tag{3.65}$$

At $r = r_0$ и $r = 6r_0 \left(\frac{r_0}{r}\right)^2 \cong 0$

$$p_{2u} = \frac{c\cos\varphi}{[\sin\varphi + 1]} \text{ and } p_{2u} = c\cot\varphi$$

Taking into account (3.57)

$$z = \frac{c}{\xi\gamma}\left(\frac{\cos\varphi}{\sin\varphi + 1}\right)$$

$$z = \frac{c}{\xi\gamma}\cot\varphi$$

The example is required to determine the maximum depth of the well, where the conditions of strength and stability of the soil. $p_2 > p_1$, $c = 20\,kPa$ $\varphi = 30^0$ and $\gamma = 16\,\kappa N/m^3$, $\xi = 0,2$

$$z = \frac{c}{\xi\gamma}\left(\frac{\cos\varphi}{\sin\varphi + 1}\right) = \frac{20}{0.2 * 16}\left(\frac{0.87}{0.5 + 1}\right) = 4\text{м}$$

$$z = \frac{c\cot\varphi}{\xi\gamma} = \frac{20 * \cot 30^0}{0,2 * 16} = \frac{35}{3,2} = 11\text{ м.}$$

i.e. at a depth of 4 m. on the surface of the wall of the well, the initial condition of strength is restored and its complete formation is completed at a depth of 11 m.

Recall that for soil located deeper than ground water it is necessary to take

$$\gamma = \frac{\gamma_s - \gamma_w}{1 + e}$$

4 Soil strength and stability of slopes, embankments and retaining walls to suit different impacts

4.1. Calculation of the stability of slopes, embankments and retaining walls considering seismic effects

In this chapter, we look at the engineering calculation methods that take into account the impact of earthquakes on the stability of slopes and retaining walls. In accordance with building regulations [23] it is assumed that areas prone to earthquakes are areas with seismic activity exceeding 7 points or more. The simplest way to take into account additional seismic loads when calculating the stability of barrier constructions and slopes is to reduce the angle of internal friction of soils by:

$$\omega = arctg\left(f_{\omega}\right) \tag{4.1}$$

where f_{ω} is the coefficient of seismic impact accounting, taken 0.04, 0.08 and 0.16, respectively, for areas with seismicity of 7.8 and 9 points. In this case, $\omega = 2.5^0$; 5^0 and 10^0.

When calculating the passive resistance of the soil, the influence of seismic effects is taken into account in accordance with the following equation

$$E_{p\omega} = \left(1 - f_{\omega}\right) E_p \tag{4.2}$$

In the design of retaining walls, we should include in the projects the following activities: massive stone retaining walls' height and length have to be reinforced with concrete cores and belts. Retaining walls along the length should be separated by seismic joints. The length of through seams should not exceed 15–20 m. The projects must be provided with free drainage outflow of groundwater outside the retaining walls.

4.2. Calculation of the stability of slopes, embankments and retaining walls considering the sagging properties of loess soils

For in situ loess soils, in calculating the stability of slopes, embankments and retaining walls, it is necessary to consider the effect of sagging properties. From the literature [18] it is known that with increasing soil moisture over perennial soils, the observed loess soils show subsidence properties, while a certain decrease in strength parameters occurs. In fig. 4.1 presents graphs of experimental results of changes in soil strength parameters with increasing moisture content.

Our studies conducted with artificially compacted soil with broken structure showed that the angle of internal friction in this case is increased by an average of 23 . . . 30%, specific adhesion decreases significantly more and approaches zero. For the loess soil of the Samarkand plateau the value of the angle of internal friction of soil compaction at the broken structure is increased by 20 . . . 25%. Obviously, this is due to increased contact between the mineral particles, and it increases the angle of internal friction of the soil. Compacted soils of irregular composition, regardless of the test procedure, have almost the same value of the angle of internal friction. In our numerous experiments with loams of disturbed soil structure of the Samarkand plateau, it was found that the value of the angle of internal friction determined by the results of single-plane shear tests ranges from 29°–31°.

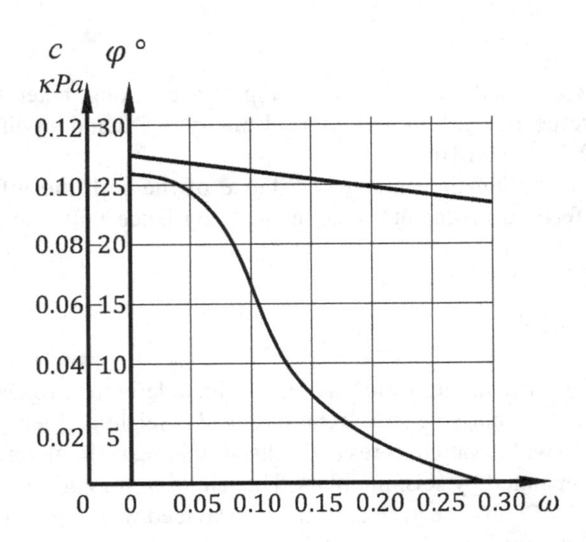

Figure 4.1 Graph of the dependence of the specific friction and the angle of internal friction on humidity.

In standard seals under water, the specific cohesion of soil undisturbed, because the destruction of the structural connections is reduced. For example, cohesion of compacted saturated soil decreases almost 10 times compared to the natural structure of the soil and humidity. In particular, it was found that for constant density with increasing humidity W_{sat} the angle of internal friction is reduced significantly by 3°, and decreases in specific cohesion by 4 to 10 times. Generalized graphs of the angle of internal friction and cohesion of specific humidity are shown in Figure 4.1.

Thus, the strength characteristics of loess soils depend on natural and structural features and, therefore, these characteristics should be determined both according to GOST (STATE STANDARD) and other methods specified in the special task for the survey.

In calculating the stability of slopes, and retaining walls it is necessary to consider possible changes in humidity and the area of their distribution. In particular, the design of retaining walls in the calculations should be based on the strength characteristics of soils corresponding to their full water saturation.

As practice shows, given the difficulties of compaction of soils in cramped conditions in the rear of the wall, often instead of these soils, the space is filled with coarse-grained sands or pebbles. In this case, this circumstance must be taken into account in the calculations, since not only the backfill material, but also the geometrical outlines of the section affect the active soil pressure on the fence. Considering that backfill of hard materials has drainage properties, there is a problem of water getting under the sole of the foundation of the retaining wall. In this case, if the soil at the base of the walls are not sealed, there are cases of deformation. For example, very often there is a slight turn of the wall, which entails filling soil subsidence and horizontal mixing of the road surface behind the retaining wall.

To determine the value of the coefficient of lateral pressure at rest, we used the device (author V.F. Sydorchuk), shown in Figure 4.2. The device consists of three side walls rigidly fixed to each other and one wall rigidly pressed to the rest by means of an adjustable clamp. Lower and upper stamps have a perforated surface for the passage of water through the soil. The housing is mounted on a compliant base and thus drastically reduces the degree of heterogeneity of the stress state in the soil sample.

Lateral pressure is measured with force cells, which are mounted on opposite walls of the device. The sensors are installed in the supporting part of the lower die, which allows controlling of the stress due to wall friction of the soil on the instrument.

When loess soil is tested on this device, it was found that the relationship between principal stresses and strains is non-linear. The coefficient of lateral pressure of loess soils at the natural humidity is 0.27 . . . 0.32, while in its

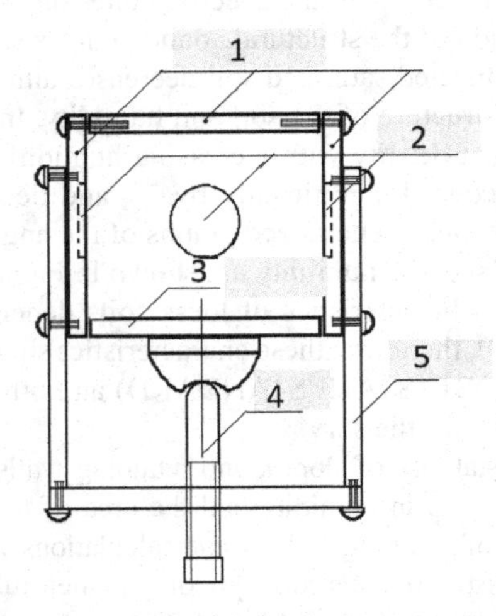

Figure 4.2 The device for determining the coefficient of lateral pressure of the soil sample in the form of a cube the size of 100 × 100 mm. 1-fixed walls; 2-force cell; 3-under-Vision panel; 4-boot screw; 5-frame.

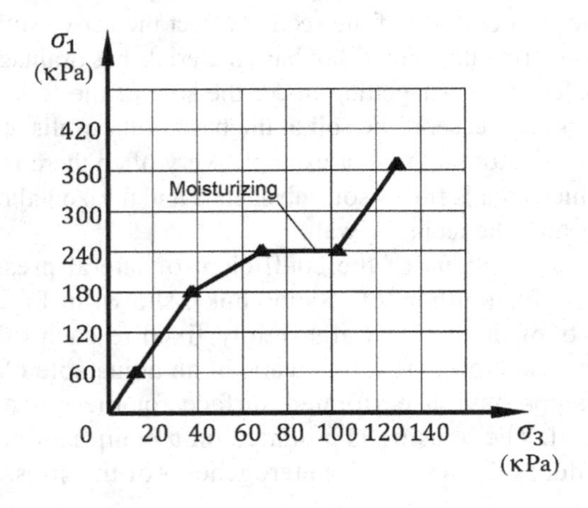

Figure 4.3 Schedule changes depending on the pressure side of the vertical load.

water-saturated state this feature is increased to 0.37 . . . 0.41. The relationship between and the principal stresses for the water-saturated loess soil has a straightforward character. This regularity is also observed in the branch of unloading the soil sample, during the soaking process with a constant value of the lateral pressure of the soil. In our tests, in the process of subsidence of loess soils, the amount of lateral pressure increased to 32 kPa (Figure 4.3).

Such a specific phenomenon of subsidence of soil must be taken into account when determining the active and passive earth pressure on the retaining wall.

Thus, based on the foregoing, we can conclude that when calculating the stability of slopes and retaining walls, it is necessary to take into account the influence of subsidence properties. In particular, values corresponding to their full moisture capacity should be introduced into the calculation of strength parameters and lateral pressure coefficient. The specific weight of the soil in the calculations should be determined taking into account the final moisture.

$$\gamma = \frac{\lambda_d}{1+W} \tag{4.4}$$

where γ_d and W – respectively, the proportion of dry weight of soil and moisture.

References

1. V.G. Berezantsev. *Asymmetric problem of the theory of limiting equilibrium loose sredy*. Moscow: Gostekhizdat, 1952.
2. A.K. Bugrov, R.N. Narbut and others. *The study of soils under triaxial*. Leningrad (Saint Petersburg): Stroyizdat, 1987, 184 pages.
3. M. Braja. *Principles of geotechnical engineering*. California State University, California, USA, 2010.
4. S.S. Golushkevich. *Static limit states mass ground*. Moscow, Russia: Gostekhizdat, 1957.
5. J.K. Zaretsky. *Lectures on modern soil mechanics*. Rostov, Russia: Rostov University, 1989.
6. G.K. Klein. *Structural mechanics loose bodies*. Moscow, Russia: Stroyizdat, 1977.
7. R.F. Craig. *Craigs soil mechanics*. London: Formerly Department of Civil Engineering University of Dundee UK, 2004.
8. G.M. Lomize and A.L. Kryzhanovsky. *On basic dependencies of VAT and soil strength: Questions strength and deformability of soils*. Baku, Azerbaijan: Baku State Institute of Civil Engineering and Architect, 1965, pp. 45–57.
9. M. Malyshev. *Soil strength and stability of the base structures*. Moscow, Russia: Stroyizdat, 1994.
10. V.V. Sokolovsky. *Static granular medium*. Moscow, Russia: Gostehizdat, 1942, 1954 and 1960.
11. A.K. Sychev and others. *Soil mechanics*. Moscow, Russia: VVITKU, 1971.
12. H.B. Seed and R.V. Witman. *Design of earth retaining structures for dynamic loads, Lateral stresses in the ground and design of earth-retaining structures*. ASCE, University of California, Richmond CA, USA, 1989, pp. 103–147.
13. D. Taylor. *The theory of soil mechanics*. Trans. from English. Moscow, Russia, 1960.
14. K. Terzaghi. *Theory of soil mechanics*. Moscow, Russia: Gosstroiizdat, 1961.
15. I. Towhata. *Geotechnical earthquake engineering*. Tokyo, Japan: Springer Verlag – Berlin, Germany, 2008. ISBN 978-3-540-35-782-7.
16. S.B. Ukhov. *Soil mechanics, foundation and bases*. Moscow, Russia: Supreme School, 2004.

17. A.Z. Hasanov and Z.A. Hasanov. *Foundations in loess subsiding soils*. Tashkent, Uzbekistan: IPTD, 2006.
18. N.A. Tsytovich. *Soil mechanics*. Moscow, Russia: Higher School, 1979.
19. D.H. Shields. Passive pressure coefficients by method of slices. *Journal of the S M and F Division, ASCE*, Vol. 99, No. Sm12. Principles of Foundation Engineering, USA, 2011.
20. The software package Plaxis 7.2 Netherlands: FAN. www.plaxis.nl.
21. KMK 2.01.03-93. Construction in seismic areas. State Commission RUz, 1996.
22. F. Tatsuoka, T. Masuda, and M.S.A. Siddiquee. Modeling the stress strain loading. *Journal Geotechnical and Environmental Engineering, ASCE*, 2002. Vol. 44, No. 2, June 2013, ISSN 0046-5828.
23. N.P. Puzyrevsky. Theory of tensions of earthy soil, L., 1929. *Collection of the Leningrad Institute of the Railway Engineers*, Vol. 95, No. 50, 1927.
24. G.S. Glushkov and V.A. Sindeev " The Material Resistance Course" published by "Vysshaya Shkola" – Moscow – 1965.

Printed in the United States
by Baker & Taylor Publisher Services

Printed in the United States
by Baker & Taylor Publisher Services